AIR-COOLED CONDENSER FUNDAMENTALS: DESIGN, OPERATIONS, TROUBLESHOOTING, MAINTENANCE, AND Q&A

CHETAN SINGH

Made with ♥ on the Notion Press Platform
www.notionpress.com

Contents

Acknowledgements

I would like to express my gratitude to all those who have supported and inspired me throughout the journey of writing this book on Air-Cooled Condenser Fundamentals: Design, Operations, Troubleshooting, Maintenance, and Q&A.

I would like to thank my mentors and colleagues in the HVAC industry for their valuable insights and expertise, which have been instrumental in shaping the content of this book. Their unwavering support and encouragement have been a source of strength and motivation for me.

I am also grateful to my family and friends who have been there for me through thick and thin, and have given me the time and space to focus on writing this book. Their love and support have been a source of inspiration for me.

Finally, I would like to express my gratitude to the readers of this book. I hope that the information contained in this book will be useful and informative, and that it will help you to better understand the complexities of air-cooled condenser systems.

Thank you all.

CHAPTER ONE

I. Introduction to Air Cooled Condensers

An air cooled condenser (ACC) is a type of heat exchanger that is used in refrigeration systems to reject heat from the refrigerant to the atmosphere. It consists of a series of tubes that are surrounded by a series of finned coils. These fins help to increase the surface area of the tubes, which allows more heat to be transferred to the surrounding air.

In an ACC system, refrigerant is circulated through the tubes, where it absorbs heat from the surrounding air. This heat is then rejected to the atmosphere through the finned coils. The ACC is typically located outside of the building, and the heat is rejected to the atmosphere through the process of convection.

One advantage of using an ACC is that it requires no water for operation, which makes it a good choice for areas where water is scarce or where the cost of water is high. ACCs are also relatively simple and inexpensive to install and maintain, which makes them a popular choice for many refrigeration systems. However, they are not as efficient as water cooled condensers in rejecting heat, and they may require more energy to operate due to the higher fan power needed to move the air.

Definition and explanation of air cooled condensers

An air cooled condenser is a type of condenser that uses air to dissipate the heat absorbed by the refrigerant during the condensation process. In a typical air cooled condenser, the refrigerant is first passed through a series of tubes or coils, which are surrounded by a series of fins. These fins help to increase the surface area of the tubes or coils, allowing them to transfer more heat to the surrounding air. The hot refrigerant gas is then passed through the tubes or coils, where it is cooled by the surrounding air and condenses into a liquid. The condensed liquid is then collected in a tray or pan and returned to the compressor to complete the refrigeration cycle.

Air cooled condensers are typically used in smaller refrigeration systems, such as those found in residential air conditioning units or small commercial refrigeration units. They are a more economical and simpler alternative to water cooled condensers, which require a separate system of pumps, pipes, and water to dissipate the heat absorbed by the refrigerant. However, air cooled condensers are less efficient than water cooled condensers, as they are not able to transfer as much heat to the surrounding air.

Air cooled condensers are typically located on the exterior of a building or on the roof of a building, as they require access to ambient air to dissipate the heat absorbed by the refrigerant. They are usually made of metal, such as aluminum or copper, and may be coated with a layer of paint or a corrosion-resistant material to protect against the elements.

Air cooled condensers are commonly used in conjunction with evaporators, which are responsible for absorbing heat from the surrounding air or liquid and transferring it to the refrigerant. The refrigerant is then passed through the condenser, where the heat is dissipated and the refrigerant is cooled and condensed. The cooled refrigerant is then returned to the evaporator to complete the refrigeration cycle.

Air cooled condensers are a common choice for smaller refrigeration systems due to their simplicity and lower cost compared to water cooled condensers. However, they are less efficient and may require more maintenance due to the exposure to outdoor elements. Water cooled condensers are a better choice for larger refrigeration systems or for applications where efficiency is a priority.

Water Cooled Condensers

Comparison to other types of heat exchangers, such as water cooled condensers.

Heat exchangers are devices that transfer heat from one fluid to another. There are many different types of heat exchangers, each with its own set of advantages and disadvantages. Some common types of heat exchangers include shell and tube heat exchangers, plate and frame heat exchangers, and air cooled heat exchangers.

Water cooled condensers are a type of heat exchanger that is used in refrigeration and air conditioning systems to transfer heat from the refrigerant to the water. Water cooled condensers are similar to shell and tube heat exchangers in that they have a tube side and a shell side, but they are specifically designed for use in refrigeration and air conditioning systems.

One advantage of water cooled condensers is that they have a high heat transfer coefficient, which means that they can transfer heat efficiently. They are also relatively compact and easy to install. However, they do require a source of cooling water, which can be expensive and may not be available in all locations.

In comparison to other types of heat exchangers, water cooled condensers may be more suitable for certain applications due to their high heat transfer coefficient and compact size. However, other types of heat exchangers may be more suitable for other applications, depending on the specific requirements of the system.

Overview of common applications and industries that use ACCs

Air cooled condensers (ACCs) are heat exchangers that are used to transfer heat from a refrigerant to the air. They are commonly used in refrigeration and air conditioning systems to reject heat from the system to the atmosphere. ACCs are a popular choice in many applications because they are relatively simple and inexpensive to install and maintain, and do not require a source of cooling water.

Some common applications and industries that use ACCs include:

- Commercial and residential air conditioning: ACCs are used in air conditioning systems to reject heat from the refrigerant to the outside air.
- Industrial cooling: ACCs are used in industrial processes to reject heat from refrigeration systems, process cooling systems, and other industrial equipment.
- Power generation: ACCs are used in power plants to reject waste heat from the generation process.
- Food and beverage processing: ACCs are used in food and beverage processing plants to reject heat from refrigeration systems used to store and transport products.
- Pharmaceutical manufacturing: ACCs are used in pharmaceutical manufacturing plants to reject heat from refrigeration and process cooling systems.
- Chemical processing: ACCs are used in chemical processing plants to reject heat from refrigeration and process cooling systems.
- Oil and gas: ACCs are used in the oil and gas industry to reject heat from refrigeration and process cooling systems used in exploration, production, and refining operations.
- HVAC: ACCs are used in heating, ventilation, and air conditioning (HVAC) systems to reject heat from the system to the outside air.

ACCs can be found in a wide range of industries and applications, and are an important component in many systems that require cooling. They are widely used due to their simplicity, low cost, and ability to reject heat effectively to the outside air.

CHAPTER TWO

II. How Air Cooled Condensers Work

Air cooled condensers (ACCs) are heat exchangers that transfer heat from a refrigerant to the air. They are commonly used in refrigeration and air conditioning systems to reject heat from the system to the atmosphere.

ACCs consist of a series of tubes through which the refrigerant flows, and a series of finned tubes or coils that are used to transfer heat from the refrigerant to the air. The fins on the tubes increase the surface area of the tubes, which allows more heat to be transferred to the air.

ACCs also have fans to blow air over the finned tubes or coils, which helps to dissipate the heat from the refrigerant to the air. The refrigerant enters the ACC at a high temperature and pressure, and as it flows through the tubes, it gives up heat to the air. The refrigerant then exits the ACC at a lower temperature and pressure, and is ready to be cooled and compressed again in the compressor.

Overall, the purpose of an ACC is to reject heat from the refrigeration or air conditioning system to the outside air. They are an important component in many systems that require cooling, and are widely used due to their simplicity, low cost, and ability to reject heat effectively to the outside air.

There are several different types of air cooled condensers, including horizontal ACCs, vertical ACCs, and split ACCs. The type of ACC used in a particular application depends on the specific requirements of the system, such as the available space, the cooling capacity needed, and the operating conditions.

Horizontal ACCs are typically used in applications where space is limited. They are mounted horizontally and have a low profile, which makes them suitable for installations with height restrictions.

Vertical ACCs are typically used in applications where there is more space available. They are mounted vertically and have a larger cooling capacity than horizontal ACCs.

Split ACCs are ACCs that are split into two or more sections, with each section having its own set of tubes and fins. Split ACCs are typically used in large systems that require a high cooling capacity.

In addition to the type of ACC, there are also several factors that can affect the performance of an ACC, including the air flow rate, the air temperature, the humidity of the air, and the refrigerant charge. Proper sizing, installation, and maintenance of the ACC are important to ensure that it operates effectively and efficiently.

Detailed explanation of the operation of ACCs

Air cooled condensers (ACCs) are heat exchangers that transfer heat from a refrigerant to the air. They are commonly used in refrigeration and air conditioning systems to reject heat from the system to the atmosphere.

The operation of an ACC involves the following steps:

- The refrigerant enters the ACC at a high temperature and pressure after being compressed in the compressor. The high temperature and pressure of the refrigerant is a result of the work done on the refrigerant by the compressor.
- As the refrigerant flows through the tubes of the ACC, it gives up heat to the air. The heat is transferred from the refrigerant to the air through a series of finned tubes or coils that are used to transfer heat from the refrigerant to the air. The fins on the tubes increase the surface area of the tubes, which allows more

heat to be transferred to the air.

- Fans are used to blow air over the finned tubes or coils, which helps to dissipate the heat from the refrigerant to the air. The air flow rate, the air temperature, and the humidity of the air all affect the performance of the ACC.
- As the refrigerant gives up heat to the air, it cools and its pressure drops. The cooled and depressurized refrigerant then flows through an expansion valve, which lowers the pressure of the refrigerant even further. The expansion valve is a flow control device that regulates the flow of refrigerant into the evaporator, where it absorbs heat from the space being cooled.
- The refrigerant then returns to the compressor, where it is compressed and the cycle begins again.
- Overall, the purpose of an ACC is to reject heat from the refrigeration or air conditioning system to the outside air. They are an important component in many systems that require cooling, and are widely used due to their simplicity, low cost, and ability to reject heat effectively to the outside air.

Description of the components of an ACC

Air cooled condensers (ACCs) are heat exchangers that transfer heat from a refrigerant to the air. They consist of several key components, including:

Tubes: The tubes of an ACC are through which the refrigerant flows. They are typically made of copper or aluminum, and are designed to withstand the high temperatures and pressures of the refrigerant.

Fins: The fins on the tubes of an ACC increase the surface area of the tubes, which allows more heat to be transferred to the air. The fins are typically made of aluminum and are attached to the tubes through a process called brazing.

Finned coils: These are coils of tubing that are surrounded by thin metal fins. The fins help to increase the surface area of the tubing, which allows more heat to be transferred to or from the refrigerant. The finned coils are used to transfer heat between the refrigerant and the air being circulated by the system.

Fans: Fans are used to blow air over the finned tubes or coils of an ACC, which helps to dissipate the heat from the refrigerant to the air. The fans are typically electric motors with blades that move the air over the fins.

Blower: The blower is a fan that is used to circulate the air through the ACC system. It is typically driven by an electric motor and is located in the ducts or near the evaporator.

Refrigerant: The refrigerant used in an ACC is a substance that absorbs and releases heat as it changes phase from a gas to a liquid and back again. Common refrigerants used in ACCs include R-22, R-410A, and R-134A.

Expansion valve: The expansion valve is a flow control device that regulates the flow of refrigerant into the evaporator, where it absorbs heat from the space being cooled. The expansion valve is located between the ACC and the evaporator, and is used to lower the pressure of the refrigerant as it enters the evaporator.

Overall, the components of an ACC work together to transfer heat from the refrigerant to the air, and to regulate the flow of refrigerant in the system. Proper sizing, installation, and maintenance of the ACC and its components are important to ensure that it operates effectively and efficiently.

In addition to the main components of an air cooled condenser (ACC), there are also several other components that may be present in an ACC system, depending on the specific design and application. These components may include:

Compressor: The compressor is a mechanical device that increases the pressure of the refrigerant by compressing it. The compressor is typically an electric motor with a piston or a screw-type mechanism that compresses the refrigerant.

Evaporator: The evaporator is a heat exchanger that absorbs heat from the space being cooled. The refrigerant flows through the evaporator and absorbs heat from the air or other fluid being cooled. As the refrigerant absorbs heat, it vaporizes and becomes a gas.

Refrigerant lines: Refrigerant lines are tubes or hoses that are used to transport the refrigerant between the various components of the ACC system. The refrigerant lines are typically made of copper, aluminum, or a high-strength plastic.

Ductwork: Ductwork is used to distribute the cooled air from the evaporator to the space being cooled. The ductwork is typically made of sheet metal or a flexible material, and is designed to withstand the pressure and temperature changes of the air being transported.

Filters: Filters are used to remove contaminants from the air being cooled by the ACC system. The filters are typically made of fiberglass or a synthetic material, and are designed to trap dust, dirt, and other particles.

Thermostat: The thermostat is a control device that monitors the temperature in the building or vehicle and adjusts the ACC system accordingly. It is typically located inside the building or vehicle and can be manually adjusted by the user to set the desired temperature.

Overall, the various components of an ACC system work together to transfer heat from the refrigerant to the air, and to circulate the refrigerant and the cooled air throughout the system. Proper sizing, installation, and maintenance of the ACC and its components are important to ensure that the system operates effectively and efficiently.

Refrigerant Heat Transfer Process

Explanation of the role of refrigerant in the heat transfer process.

Refrigerant is a substance that is used in a heat pump or air conditioning system to transfer heat from one place to another. It is used to absorb heat from the air or liquid in one location and release it in another.

In a typical air conditioning system, the refrigerant is responsible for absorbing heat from the air inside of a building and transferring it to the outdoor air. It does this by moving through a series of tubes and coils that make up the system.

The refrigerant starts off in a low-pressure, low-temperature liquid form. As it passes through the system, it absorbs heat and becomes a high-pressure, high-temperature gas. The gas is then condensed back into a liquid, releasing the heat that it absorbed. This process is repeated over and over, allowing the air conditioning system to continuously remove heat from the air inside of a building and transfer it outside.

Overall, the role of the refrigerant in the heat transfer process is to absorb heat from one location and release it in another, helping to regulate the temperature and humidity of a space.

CHAPTER THREE

III. Advantages and Disadvantages of Air Cooled Condensers

Air cooled condensers are a type of condenser that are used in refrigeration and air conditioning systems to reject heat that has been absorbed during the cooling process. They use ambient air as a cooling medium, rather than water, to cool the refrigerant that is being condensed.

Advantages of air cooled condensers include:

- Easy to install: Air cooled condensers do not require any special piping or water connections, which makes them relatively easy to install.
- Low cost: Air cooled condensers are generally less expensive to install and operate than water cooled condensers, since they do not require a water supply or treatment system.
- Low maintenance: Air cooled condensers do not require as much maintenance as water cooled condensers, since they do not have to be treated for scale or corrosion.
- Can be used in areas with limited water availability: Air cooled condensers are a good choice for areas where water is scarce or where there are water usage restrictions.

Disadvantages of air cooled condensers include:

- Lower cooling efficiency: Air cooled condensers are generally less efficient at rejecting heat than water cooled condensers, since the cooling capacity of air is lower than that of water.
- Limited by ambient air temperature: The performance of an air cooled condenser can be affected by high ambient air temperatures, since the air used to cool the refrigerant may be warmer and less effective.
- May require additional space: Air cooled condensers may require more space for installation, since they typically have larger fins or coils to increase their surface area for heat rejection.
- Overall, the choice between an air cooled condenser and a water cooled condenser will depend on the specific needs and constraints of a particular application.

Discussion of the benefits of ACCs, including low cost and water conservation

Air cooled condensers (ACCs) are a type of condenser that are used in refrigeration and air conditioning systems to reject heat that has been absorbed during the cooling process. They use ambient air as a cooling medium, rather than water, to cool the refrigerant that is being condensed. ACCs have several benefits that make them a popular choice in a variety of applications.

One of the main benefits of ACCs is their low cost. They are generally less expensive to install and operate than water cooled condensers, since they do not require a water supply or treatment system. This can be especially important in areas where water is scarce or where there are water usage restrictions.

Another benefit of ACCs is their ability to conserve water. Water cooled condensers require a constant supply of water to operate, which can be a problem in areas with limited water resources or during times of drought. ACCs, on the other hand, do not use any water in the cooling process, making them a more environmentally friendly and sustainable choice.

ACCs are also relatively easy to install, since they do not require any special piping or water connections. They are also low maintenance, since they do not have to be treated for scale or corrosion like water cooled condensers do.

While ACCs do have some limitations, such as lower cooling efficiency and sensitivity to ambient air temperature, they can be a good choice for many applications where the benefits of low cost and water conservation are important.

Comparison of ACC efficiency to other types of heat exchangers

Air cooled condensers (ACCs) are a type of heat exchanger that is used to reject heat from a refrigeration or air conditioning system. They use ambient air as a cooling medium to cool the refrigerant that is being condensed. The efficiency of an ACC is typically lower than that of other types of heat exchangers, such as water cooled condensers or evaporative condensers.

One reason for this is that the cooling capacity of air is lower than that of water. Water has a higher specific heat capacity, which means that it can absorb more heat per unit of mass than air can. As a result, water cooled condensers are generally more efficient at rejecting heat than ACCs.

Another factor that can impact the efficiency of an ACC is ambient air temperature. If the air temperature is high, the ACC will be less effective at cooling the refrigerant, leading to lower overall efficiency.

Evaporative condensers are another type of heat exchanger that can be more efficient than ACCs, especially in warm, dry climates. These condensers use water and a natural draft to cool the refrigerant, and can be more effective at rejecting heat than ACCs under certain conditions.

Overall, the choice between an ACC and another type of heat exchanger will depend on the specific needs and constraints of a particular application. Factors to consider include the available cooling media, the climate, and the desired level of efficiency.

Examination of potential drawbacks of ACCs, such as higher energy consumption and potential for noise pollution

Air cooled condensers (ACCs) are a type of heat exchanger that is used to reject heat from a refrigeration or air conditioning system. They use ambient air as a cooling medium to cool the refrigerant that is being condensed. While ACCs have several benefits, such as low cost and ease of installation, they also have some potential drawbacks that should be considered.

One potential drawback of ACCs is higher energy consumption. ACCs are generally less efficient at rejecting heat than other types of heat exchangers, such as water cooled condensers or evaporative condensers. This means that they may require more energy to operate, which can lead to higher operating costs.

Another potential drawback of ACCs is the potential for noise pollution. ACCs typically use fans to blow ambient air over the fins or coils that are used to reject heat. These fans can generate significant noise, which can be a problem in residential or commercial settings where noise levels need to be kept to a minimum.

ACCs may also be less effective at rejecting heat in high ambient air temperatures, which can limit their performance and lead to reduced efficiency.

Overall, while ACCs have some potential drawbacks, they can be a good choice in many applications where the benefits of low cost and ease of installation outweigh these potential issues. It is important to carefully consider the specific needs and constraints of a particular application when deciding whether to use an ACC.

CHAPTER FOUR

IV. Designing ACC System Factors

Overview of factors to consider when designing an ACC air cooled condenser system, including climate, system size, and cost.

There are several factors to consider when designing an air cooled condenser (ACC) system:

- **Climate:** The ambient temperature and humidity of the location where the ACC system will be installed can have a significant impact on its performance. For example, an ACC system may not be as effective in a hot and humid climate as it would be in a dry climate.
- **System size:** The size of the ACC system should be chosen based on the cooling load of the facility. It is important to accurately calculate the cooling load to ensure that the ACC system is properly sized.
- **Cost:** The cost of an ACC system will depend on its size, as well as the materials used and the complexity of the design. It is important to carefully consider the trade-offs between initial cost and long-term energy efficiency.
- **Maintenance requirements:** ACC systems require regular maintenance, such as cleaning the air filters and inspecting the fans and coils, to ensure that they are operating efficiently. It is important to consider the maintenance requirements when designing an ACC system.

- **Noise:** ACC systems can be noisy, especially if they are located near residential or sensitive areas. It is important to consider the noise level of the ACC system and take steps to mitigate it if necessary.
- **Safety:** ACC systems should be designed with safety in mind, including provisions for emergency shutdown and the use of safety guards and barriers to protect workers.

Explanation of the process of selecting an ACC air cooled condenser based on these factors

The process of selecting an air cooled condenser (ACC) system based on the above factors might involve the following steps:

- **Determine the cooling load:** The first step in selecting an ACC system is to accurately calculate the cooling load of the facility. This will help to ensure that the system is properly sized to meet the cooling needs of the facility.
- **Consider the climate:** The ambient temperature and humidity of the location where the ACC system will be installed should be taken into account when selecting the system. A system that is well-suited to a dry climate may not perform as well in a hot and humid climate.
- **Evaluate cost:** The cost of the ACC system should be considered, including the initial cost of the system as well as ongoing maintenance and energy costs. It may be worth paying a higher initial cost for a more energy-efficient system if it will result in lower operating costs over the long term.
- **Consider maintenance requirements:** ACC systems require regular maintenance to ensure that they are operating efficiently. It is important to consider the maintenance requirements of different systems and choose one that is appropriate for the facility.

- **Evaluate noise levels:** ACC systems can be noisy, especially if they are located near residential or sensitive areas. It is important to consider the noise levels of different systems and choose one that is appropriate for the location.
- **Consider safety:** ACC systems should be designed with safety in mind, including provisions for emergency shutdown and the use of safety guards and barriers to protect workers. It is important to choose a system that meets all relevant safety standards.

Discussion of ACC sizing and capacity requirements

The size of an air cooled condenser (ACC) system is an important factor to consider when designing a cooling system. An ACC system that is too small will not be able to meet the cooling needs of the facility, while an ACC system that is too large will be inefficient and may waste energy.

To determine the proper size of an ACC system, it is necessary to calculate the cooling load of the facility. The cooling load is a measure of the amount of heat that must be removed from a building or facility in order to maintain a comfortable temperature. It is typically expressed in BTUs per hour (BTU/hr).

There are several factors that can affect the cooling load of a facility, including the size of the facility, the type of insulation used, the number of windows and doors, and the local climate. Accurately calculating the cooling load is essential in order to properly size the ACC system.

Once the cooling load has been determined, the size of the ACC system can be chosen based on the capacity required to meet the cooling load. ACC systems are typically rated in terms of their cooling capacity, which is typically expressed in tons. One ton of cooling capacity is equivalent to the ability to remove 12,000 BTUs of heat per hour.

It is important to select an ACC system with a capacity that is appropriate for the cooling load of the facility. An ACC system that is too small will not be able to meet the cooling needs of the facility, while an ACC system that is too large will be inefficient and may waste energy.

CHAPTER FIVE

V. Air cooled condenser ACC Maintenance and Troubleshooting

Proper maintenance is essential for ensuring that an air cooled condenser (ACC) system is operating efficiently and effectively. Some common maintenance tasks for ACC systems include:

- Cleaning air filters: Air filters should be regularly cleaned or replaced to ensure that air can flow freely through the ACC system.
- Inspecting fans and coils: The fans and coils of an ACC system should be regularly inspected for damage or debris. Any damage or debris should be repaired or removed to ensure that the system is operating effectively.
- Checking refrigerant levels: The refrigerant levels in an ACC system should be checked regularly to ensure that the system is properly charged.
- Lubricating bearings: The bearings of the fans in an ACC system should be lubricated regularly to ensure that they are operating smoothly.

If an ACC system is not functioning properly, there are several steps that can be taken to troubleshoot the issue:

- Check the power supply: Make sure that the ACC system is properly connected to a power source and that all circuit breakers and fuses are functioning properly.
- Check the thermostat: Make sure that the thermostat is set to the correct temperature and that it is functioning properly.
- Check the air filters: Dirty or clogged air filters can restrict airflow and cause the ACC system to perform poorly.
- Check the refrigerant levels: Low refrigerant levels can cause the ACC system to underperform.
- Check for leaks: Leaks in the ACC system can cause the system to underperform or malfunction.
- If the issue cannot be resolved through these steps, it may be necessary to call a professional technician for further assistance.

Explanation of the importance of regular ACC maintenance

Regular maintenance is important for ensuring that an air cooled condenser (ACC) system is operating efficiently and effectively. Some of the benefits of regular ACC maintenance include:

- Improved efficiency: Regular maintenance can help to ensure that the ACC system is operating at peak efficiency, which can help to reduce energy costs.
- Increased lifespan: Proper maintenance can help to extend the lifespan of the ACC system by preventing wear and tear on the system's components.
- Enhanced performance: Regular maintenance can help to improve the performance of the ACC system, ensuring that it is able to effectively cool the facility.
- Early detection of problems: Regular maintenance can help to identify potential problems with the ACC system before they become serious issues. This can help to prevent costly repairs and downtime.

- Compliance with warranties: Many ACC systems come with warranties that require regular maintenance in order to remain valid. Failing to perform regular maintenance can void the warranty and leave the facility vulnerable to costly repairs.

Overall, regular ACC maintenance is an important aspect of ensuring that the system is operating effectively and efficiently. It is important to follow the manufacturer's recommended maintenance schedule and to perform regular inspections and cleaning to keep the system in good working order.

Description of common ACC problems and troubleshooting techniques

There are several common problems that can occur with air cooled condenser (ACC) systems, including:

- Insufficient cooling: This can be caused by a variety of factors, such as low refrigerant levels, dirty or clogged air filters, or malfunctioning thermostats.
- Loud noises: ACC systems may make loud noises if there is a problem with the fans or the bearings are not properly lubricated.
- Leaks: Leaks in the ACC system can cause the system to underperform or malfunction.
- Frozen coils: Frozen coils can occur if the ACC system is low on refrigerant or if there is a problem with the airflow.
- If an ACC system is experiencing any of these problems, there are several steps that can be taken to troubleshoot the issue:
- Check the power supply: Make sure that the ACC system is properly connected to a power source and that all circuit breakers and fuses are functioning properly.
- Check the thermostat: Make sure that the thermostat is set to the correct temperature and that it is functioning properly.

- Check the air filters: Dirty or clogged air filters can restrict airflow and cause the ACC system to perform poorly.
- Check the refrigerant levels: Low refrigerant levels can cause the ACC system to underperform.
- Check for leaks: Leaks in the ACC system can cause the system to underperform or malfunction.

If the issue cannot be resolved through these steps, it may be necessary to call a professional technician for further assistance.

Discussion of ACC repair and replacement options

If an air cooled condenser (ACC) system is experiencing problems and cannot be repaired, there are several options for replacing the system. Some options include:

- Repairing the ACC system: If the issue with the ACC system is minor, it may be possible to repair the system rather than replacing it. This can be a cost-effective option if the system is relatively new or if the repair is relatively simple.
- Replacing the ACC system: If the ACC system is old or if the repair would be costly, it may be more cost-effective to replace the system. When replacing an ACC system, it is important to consider the size and capacity of the system, as well as the overall cost of the replacement.
- Upgrading the ACC system: In some cases, it may be possible to upgrade the ACC system rather than replacing it. This can involve adding new features or components to the system to improve its performance or efficiency.
- Retrofitting the ACC system: Retrofitting an ACC system involves making modifications to the system to improve its performance or efficiency. This can be a cost-effective option if the system is relatively new and in good condition.

When deciding on a course of action, it is important to consider the age of the ACC system, the cost of repair or replacement, and the potential energy savings that can be achieved through a new or upgraded system.

CHAPTER SIX

VI Technology

What is Future Trends in Air Cooled Condenser Technology.

Air cooled condensers are a type of heat exchanger that are used to cool refrigerants or other fluids by dissipating heat to the air. Here are some possible trends in air cooled condenser technology:

- Increased use of low-global warming potential (GWP) refrigerants: As concerns about climate change and environmental sustainability continue to grow, there is increasing pressure to use refrigerants with low GWP in air cooled condensers. Low-GWP refrigerants have a lower potential to contribute to global warming when they are released into the atmosphere.
- Improved energy efficiency: There is a growing demand for more energy-efficient air cooled condensers, as energy costs continue to rise and as consumers and businesses become more environmentally conscious. Manufacturers are working on developing more efficient designs and materials to reduce the amount of energy required to operate air cooled condensers.
- Use of advanced materials: Researchers are exploring the use of advanced materials, such as metals and alloys with high thermal conductivity, in the construction of air cooled condensers. These materials have the potential to improve the heat transfer efficiency of the condenser and reduce energy consumption.
- Development of microchannel technology: Microchannel technology involves the use of small channels, typically less than

1 mm in diameter, to increase the surface area available for heat transfer. This can potentially improve the efficiency of air cooled condensers, especially in small-scale applications.

- Increasing use of variable speed drives: Variable speed drives (VSDs) allow the speed of the air cooled condenser's fan to be adjusted based on the cooling load. This can improve the energy efficiency of the condenser by reducing the amount of energy required to operate the fan.

Overview of advances in ACC design and materials

Here are some key advances in the design and materials used in air cooled condensers (ACCs):

- Microchannel technology: As mentioned earlier, microchannel technology involves the use of small channels to increase the surface area available for heat transfer. This can potentially improve the efficiency of ACCs, especially in small-scale applications.
- Use of advanced materials: Researchers are exploring the use of advanced materials, such as metals and alloys with high thermal conductivity, in the construction of ACCs. These materials have the potential to improve the heat transfer efficiency of the condenser and reduce energy consumption.
- Optimized fin design: The fins on an ACC are responsible for dissipating heat to the surrounding air. Manufacturers are working on optimizing the design of these fins to improve the heat transfer efficiency of the ACC.
- Use of variable speed drives: Variable speed drives (VSDs) allow the speed of the ACC's fan to be adjusted based on the cooling load. This can improve the energy efficiency of the ACC by reducing the amount of energy required to operate the fan.
- Improved air flow design: The design of the air flow path through an ACC can have a significant impact on its

performance. Manufacturers are working on optimizing the design of the air flow path to improve the heat transfer efficiency of the ACC.

- Use of low-GWP refrigerants: As concerns about climate change and environmental sustainability continue to grow, there is increasing pressure to use refrigerants with low global warming potential (GWP) in ACCs. Low-GWP refrigerants have a lower potential to contribute to global warming when they are released into the atmosphere.

Examination of potential developments in ACC energy efficiency and performance

There are several potential developments that could improve the energy efficiency and performance of air cooled condensers (ACCs):

- Use of advanced materials: Researchers are exploring the use of advanced materials, such as metals and alloys with high thermal conductivity, in the construction of ACCs. These materials have the potential to improve the heat transfer efficiency of the condenser and reduce energy consumption.
- Improved fin design: The fins on an ACC are responsible for dissipating heat to the surrounding air. Manufacturers are working on optimizing the design of these fins to improve the heat transfer efficiency of the ACC.
- Use of variable speed drives: Variable speed drives (VSDs) allow the speed of the ACC's fan to be adjusted based on the cooling load. This can improve the energy efficiency of the ACC by reducing the amount of energy required to operate the fan.
- Improved air flow design: The design of the air flow path through an ACC can have a significant impact on its performance. Manufacturers are working on optimizing the design of the air flow path to improve the heat transfer

efficiency of the ACC.

- Use of low-GWP refrigerants: As concerns about climate change and environmental sustainability continue to grow, there is increasing pressure to use refrigerants with low global warming potential (GWP) in ACCs. Low-GWP refrigerants have a lower potential to contribute to global warming when they are released into the atmosphere.
- Development of microchannel technology: Microchannel technology involves the use of small channels, typically less than 1 mm in diameter, to increase the surface area available for heat transfer. This can potentially improve the efficiency of ACCs, especially in small-scale applications.
- Use of smart controls: The use of smart controls, such as building management systems or internet of things (IoT) devices, can improve the energy efficiency of ACCs by allowing them to be controlled and monitored remotely. This can enable ACCs to be operated more efficiently and reduce energy consumption.

Discussion of the role of ACCs in the wider context of sustainable and renewable energy systems

Air cooled condensers (ACCs) play a role in the wider context of sustainable and renewable energy systems in a number of ways:

- ACCs are used in a variety of systems that rely on refrigeration or air conditioning, including refrigerated warehouses, supermarkets, and air conditioning systems in buildings. These systems can be powered by renewable energy sources, such as solar or wind power, which can reduce their environmental impact and contribute to the adoption of sustainable energy.
- ACCs can be used to recover waste heat from industrial processes, such as those in the petrochemical or power generation industries. This can increase the overall efficiency of

the process and reduce the amount of energy that is wasted.

- ACCs can be designed to be more energy efficient, which can reduce the overall energy consumption of systems that use them. This can help to reduce greenhouse gas emissions and contribute to the transition to a more sustainable energy system.
- ACCs can be used in conjunction with other sustainable technologies, such as ground source heat pumps or solar thermal systems, to provide an efficient and environmentally friendly means of heating and cooling buildings.
- The use of low-global warming potential (GWP) refrigerants in ACCs can also contribute to the sustainability of energy systems by reducing the potential for these refrigerants to contribute to global warming when they are released into the atmosphere.

CHAPTER SEVEN

VII. Pros and cons of water-cooled and air-cooled condensers for HVAC systems

Pros of Water-Cooled Condensers:

- Increased Efficiency: Water-cooled condensers are generally more efficient at removing heat compared to air-cooled condensers, due to the higher heat transfer capacity of water.
- Space Savings: Water-cooled condensers can take up less space compared to air-cooled condensers, as they can be located outside of the building or in a smaller, more compact area.
- Reduced Noise: Water-cooled condensers are generally quieter than air-cooled condensers, as they do not require the use of large fans.
- Water Conservation: In regions where water is scarce, water-cooled condensers can be designed to conserve water, as they can use recirculated water instead of new water.

Cons of Water-Cooled Condensers:

- Increased Complexity: Water-cooled condensers are generally more complex to install and maintain compared to air-cooled condensers, due to the need for a separate water system and the potential for leaks and other issues.
- Higher Initial Cost: The initial cost of water-cooled condensers is typically higher than that of air-cooled condensers, due to the increased complexity and additional components required.
- Water Availability: The availability of a sufficient supply of cooling water is a critical requirement for water-cooled condensers, whereas air-cooled condensers do not require a separate water supply.
- Water Quality: The quality of the cooling water must be carefully monitored to prevent corrosion, scaling, and other issues that can impact the performance and longevity of the condenser.
- Water Disposal: The discharge of cooling water from water-cooled condensers must be managed in an environmentally responsible manner, which can add additional costs and challenges.

Pros of Air-Cooled Condensers:

- Simplicity: Air-cooled condensers are generally simpler to install and maintain compared to water-cooled condensers, as they do not require a separate water system.
- Lower Initial Cost: The initial cost of air-cooled condensers is typically lower than that of water-cooled condensers, due to the simplicity of the design and fewer components required.
- Flexibility: Air-cooled condensers can be located anywhere, as they do not require a separate water supply.

Cons of Air-Cooled Condensers:

- Reduced Efficiency: Air-cooled condensers are generally less efficient at removing heat compared to water-cooled condensers, due to the lower heat transfer capacity of air.

- Increased Noise: Air-cooled condensers can be noisy, as they require the use of large fans to circulate air.
- Space Requirements: Air-cooled condensers can take up more space compared to water-cooled condensers, as they require a large area for air flow and the use of fans.

CHAPTER EIGHT

VIII. Latest air-cooled condenser technologies

The latest technologies in air-cooled condenser systems include variable speed fans, electronic expansion valves, and microchannel condenser coils.

- Variable Speed Fans: Variable speed fans are a modern technology in air-cooled condenser systems that allows for adjustable fan speeds. By controlling the fan speed, the system can match the cooling needs with the available capacity, leading to improved energy efficiency and reduced operating costs.
- Electronic Expansion Valves: Electronic expansion valves are a type of metering device used in air-cooled condenser systems to control the flow of refrigerant. The valve opens or closes based on a signal from the control system, ensuring that the right amount of refrigerant is flowing through the system. This improved control leads to better efficiency, reduced energy consumption, and longer equipment life.
- Microchannel Condenser Coils: Microchannel condenser coils are a type of heat exchanger used in air-cooled condenser systems. The coils are made up of small, flat tubes with a high surface area-to-volume ratio. This increased surface area allows for more efficient heat transfer, leading to improved energy efficiency and reduced operating costs. The compact design of microchannel condenser coils also reduces the size and weight

of the system, making it easier to install and maintain.

These latest technologies in air-cooled condenser systems are driving improvements in efficiency, reliability, and cost-effectiveness. By adopting these technologies, air-cooled condenser systems can better meet the demands of modern refrigeration systems while reducing their environmental impact.

CHAPTER NINE

IX. Air cooled condenser related Q&A

Here are a few questions and answers about air cooled condensers:

What is an air cooled condenser?

An air cooled condenser is a type of heat exchanger that is used to cool a refrigerant or other fluid by transferring heat from the fluid to the air. It consists of a series of tubes or coils through which the fluid flows, and fins attached to the tubes or coils to increase the surface area for heat transfer.

How does an air cooled condenser work?

An air cooled condenser works by using a fan to blow air over the tubes or coils, which cools the refrigerant or other fluid inside. As the air passes over the tubes or coils, it picks up heat from the fluid and carries it away, lowering the temperature of the fluid. The cooled fluid is then returned to the system to absorb more heat, and the process repeats.

What are the advantages of an air cooled condenser?

Some advantages of air cooled condensers include their compact size, low maintenance requirements, and the fact that they do not require a water source for cooling. They are also relatively simple and inexpensive to install, and can be used in a wide range of applications.

What are the limitations of an air cooled condenser?

One potential limitation of air cooled condensers is that they are not as efficient at removing heat as water cooled condensers, which use water to transfer heat away from the system. Additionally, they may be less effective in hot, humid environments, as the air may be less able to absorb heat due to the high humidity.

When is an air cooled condenser used?

Air cooled condensers are often used in applications where a water source is not available or where the cost of installing a water cooled system would be prohibitive. They are commonly used in refrigeration and air conditioning systems, as well as in power plants and other industrial processes.

What are the advantages of an air-cooled condenser in a thermal power plant?

There are several advantages of using air-cooled condensers in a thermal power plant:

- Water Conservation: Air-cooled condensers do not require water for cooling, unlike traditional water-cooled condensers, reducing the demand for water in a thermal power plant and helping to conserve this precious resource.
- Improved Efficiency: Air-cooled condensers typically have a smaller footprint than water-cooled condensers, which results in improved plant efficiency. The smaller size allows for more compact design and reduced material costs.
- Lower Operating Costs: Air-cooled condensers do not require water treatment, which helps reduce the operating costs associated with water-cooled condensers. The lack of water also eliminates the need for cooling towers and pumps, reducing the maintenance and repair costs associated with these components.
- Reduced Environmental Impact: Air-cooled condensers do not discharge warm water into the environment, reducing the environmental impact associated with water-cooled condensers.
- Increased Reliability: Air-cooled condensers do not have water-side issues such as corrosion, scaling, or fouling, which can occur in water-cooled condensers. This increased reliability results in improved plant availability and reduced downtime etc.

Is that necessary to use condenser in thermal power plant? Why?

- Yes, it is necessary to use a condenser in a thermal power plant. The condenser plays a critical role in the operation of a steam turbine, which is a common power generation technology used in thermal power plants.
- The condenser's primary function is to condense the steam that has passed through the turbine back into water. This process is necessary because the condensed water can be reused to generate more steam, increasing the efficiency of the power plant.
- Additionally, the condenser helps to maintain a vacuum in the steam turbine, which is necessary for efficient operation. The vacuum in the steam turbine helps to reduce energy losses, increase the turbine's output, and reduce the risk of damage to the turbine components.

In conclusion, the use of a condenser in a thermal power plant is necessary because it helps to increase the efficiency and reliability of the steam turbine, which is a critical component in power generation.

What happens when a thermal power plant condenser vacuum is high?

When a thermal power plant condenser's vacuum is high, it indicates that the pressure in the condenser is lower than the atmospheric pressure. This high vacuum level is important for the efficient operation of the steam turbine in the power plant.

A high vacuum level in the condenser results in several benefits, including:

- Increased efficiency: The high vacuum level reduces the amount of energy required to condense the steam back into water, which increases the overall efficiency of the power plant.

- Reduced energy losses: The high vacuum level reduces energy losses in the steam turbine, which results in increased output and better performance.
- Improved turbine performance: The high vacuum level helps to prevent the steam from condensing in the turbine, which can cause damage to the turbine components.
- Increased reliability: The high vacuum level in the condenser helps to maintain the efficiency and reliability of the steam turbine, reducing the risk of downtime and maintenance requirements.

However, if the vacuum level in the condenser is too high, it can result in excessive energy losses and reduced efficiency. It is important to maintain a balance between the vacuum level and the pressure in the steam turbine to ensure optimal performance and efficiency.

Can an existing power plant with a water-cooled condenser be converted to an air-cooled condenser?

Yes, it is possible to convert a power plant with a water-cooled condenser to an air-cooled condenser, although it may require significant modifications to the plant's infrastructure and equipment. The conversion process may involve removing the water-cooled system and replacing it with an air-cooling system, which may require changes to the plant's layout, equipment and cooling tower. The feasibility of the conversion will depend on various factors, such as the local climate, water availability, and regulations, and it is advisable to seek expert advice to evaluate the potential cost, technical and regulatory challenges involved.

How does an air-cooled condenser work in a steam power plant?

In a steam power plant, an air-cooled condenser (ACC) replaces the traditional water-cooled condenser to condense the steam produced by the steam turbine. The ACC works by using air as the cooling medium instead of water to remove the heat from the condensed steam.

Here's how it works:

- The hot, high-pressure steam from the steam turbine is directed into the ACC, where it flows over a series of finned tubes.
- The heat from the steam is transferred to the finned tubes, which are cooled by air that is drawn over them by fans.
- The cooled steam is then condensed into water, which is collected and sent back to the steam generator for reuse.
- The hot air generated by the cooling process is exhausted to the atmosphere.

In an air-cooled condenser, there is no need for a cooling tower or a large water supply, making it a more cost-effective and water-saving solution for power plants located in areas with limited water resources or high evaporation rates. However, the performance of ACCs may be limited by the local climate and ambient temperature, as high temperatures can decrease their efficiency and increase the cooling load required.

How does wind direction affect air-cooled condenser (ACC) orientation?

- The wind direction can affect the performance of an air-cooled condenser (ACC) by influencing the flow of air over the finned tubes. The optimal orientation of the ACC is one that allows for the most efficient flow of air over the tubes, which can help to maximize the heat transfer rate and improve the condensation process.
- In general, ACCs are designed to be oriented so that the wind direction is perpendicular to the direction of the finned tubes. This allows for the maximum amount of air to pass over the tubes and helps to ensure that the ACC operates at peak efficiency. If the wind direction is not perpendicular to the tubes, the flow of air over the tubes may be reduced, which can reduce the efficiency of the condenser.
- It is important to note that the wind direction may vary throughout the day and can be influenced by local weather patterns and conditions. As a result, some ACC designs

incorporate adjustable fins or louvers that can be adjusted to optimize the flow of air over the tubes based on the prevailing wind direction. In other cases, the ACC may be designed with multiple sections that can be independently adjusted based on the wind direction.

- In general, the orientation of the ACC should be carefully considered during the design phase of the power plant, and the performance of the ACC should be monitored and evaluated regularly to ensure that it is operating at peak efficiency.

What are the disadvantages of using an air cooled condenser?

There are several disadvantages of using an air-cooled condenser (ACC) in a steam power plant, including:

- Decreased efficiency: ACCs are typically less efficient than water-cooled condensers, especially in hot climates. The high ambient temperature can reduce the heat transfer rate and increase the cooling load required, reducing the overall efficiency of the power plant.
- Increased power consumption: ACCs require large fans to circulate air over the finned tubes, which increases the power consumption of the power plant. This additional energy consumption can increase the operating costs of the power plant.
- Higher capital costs: The cost of purchasing and installing an ACC is typically higher than that of a water-cooled condenser due to the need for large fans and specialized equipment.
- Limited capacity: ACCs have a limited capacity to condense steam, so they are typically used in smaller power plants or in combination with water-cooled condensers.
- Increased land usage: ACCs require a large area of land to be installed, as they require significant space for the fans and finned tubes. This can limit their use in densely populated areas.
- Higher noise levels: ACCs generate significant noise due to the operation of the large fans, which can be a nuisance to nearby

communities.

In general, the suitability of ACCs for a specific power plant will depend on several factors, including the local climate, water availability, regulations, and cost. It is important to carefully consider the advantages and disadvantages of ACCs when evaluating the potential cooling options for a power plant.

What are the advantages of using water cooled condensers in power plants over air cooled condensers?

There are several advantages of using water-cooled condensers (WCCs) over air-cooled condensers (ACCs) in power plants, including:

- Increased efficiency: WCCs are typically more efficient than ACCs, especially in cool climates, as the lower ambient temperature allows for more efficient heat transfer from the condensed steam.
- Lower energy consumption: WCCs do not require large fans to circulate air over the tubes, which reduces the power consumption of the power plant compared to ACCs.
- Lower capital costs: The cost of purchasing and installing a WCC is typically lower than that of an ACC, as they require less specialized equipment.
- Higher capacity: WCCs have a higher capacity to condense steam compared to ACCs, making them more suitable for large power plants.
- Reduced land usage: WCCs require less land compared to ACCs, as they do not require large fans or a significant amount of space for the finned tubes.
- Lower noise levels: WCCs generate less noise compared to ACCs, as they do not require the operation of large fans.

In general, the suitability of WCCs for a specific power plant will depend on several factors, including the local climate, water availability, regulations, and cost. It is important to carefully consider the advantages and disadvantages of both WCCs and ACCs when evaluating the potential cooling options for a power plant.

Water cooled condenser and air-cooled condenser what is the impact in summer and cold condition, which one is more efficient rather than cost?

- The efficiency of water-cooled condensers (WCCs) and air-cooled condensers (ACCs) can be affected by the local weather conditions, specifically temperature and humidity.
- In summer, the high ambient temperature can reduce the efficiency of ACCs, as the increased cooling load required can reduce the heat transfer rate. In contrast, WCCs can maintain their efficiency as they use a continuous supply of cool water to condense the steam. However, the availability of water during periods of drought or water scarcity can limit the use of WCCs in some regions.
- In cold weather, the efficiency of WCCs can be reduced due to the lower temperature of the cooling water, which can reduce the heat transfer rate. In contrast, ACCs can maintain their efficiency as the ambient temperature is typically higher than the cooling water temperature used in WCCs.
- In terms of cost, the initial capital cost of installing an ACC is typically lower than that of a WCC. However, the operating costs of ACCs are typically higher, as they require large fans to circulate air over the finned tubes, which increases the power consumption of the power plant. In contrast, WCCs have lower operating costs as they do not require large fans.

In conclusion, the most efficient and cost-effective option between WCCs and ACCs will depend on several factors, including the local climate, water availability, regulations, and specific requirements of the power plant. It is important to carefully consider these factors when evaluating the cooling options for a power plant.

How does an ejector work in an air cooled condenser turbine and create a vacuum?

An ejector in an air-cooled condenser (ACC) turbine is a device used to create a vacuum by using the kinetic energy of a high-pressure fluid to entrain and mix with a low-pressure fluid. In the case of an ACC turbine, the high-pressure fluid is usually steam, while the low-pressure fluid is the air used to cool the ACC.

Here's how an ejector in an ACC turbine works:

- Steam is supplied to the nozzle of the ejector from the high-pressure side of the turbine.
- The high-speed steam from the nozzle entrains the air from the ACC, mixing it with the steam to form a mixture.
- The mixture then flows into the mixing chamber where it slows down and releases its kinetic energy.
- As the mixture slows down, the pressure in the mixing chamber decreases, creating a vacuum.
- The vacuum in the mixing chamber pulls in additional air from the ACC, which is then mixed with the steam to form a new mixture.
- The process continues, creating a continuous flow of air and steam through the ejector, which maintains the vacuum in the ACC.
- The vacuum created by the ejector in the ACC helps to increase the efficiency of the power plant by reducing the back pressure on the turbine and increasing the amount of steam that can be condensed. The ejector also helps to remove non-condensable gases, such as air and nitrogen, from the ACC, which can improve the heat transfer rate and reduce the cooling load required.

In air cooled condensers, why are fin-tube type heat exchangers used?

In air-cooled condensers (ACCs), fin-tube type heat exchangers are commonly used to provide efficient heat transfer from the steam to the surrounding air. The fin-tube heat exchanger consists of tubes, typically made of copper or aluminum, that carry the steam, and fins, usually made of aluminum, that increase the surface area for heat transfer to the air.

The use of fin-tube heat exchangers in ACCs provides several benefits, including:

- Increased heat transfer surface area: The fins on the tubes increase the surface area for heat transfer, which allows for efficient heat transfer from the steam to the air.
- Improved air flow: The fins create turbulence in the air flow, which increases the heat transfer rate.
- Reduced air resistance: The fins on the tubes break up the air flow and reduce air resistance, which allows for a more efficient cooling process.
- Improved durability: The tubes and fins are made of durable materials, such as copper and aluminum, which can withstand the harsh conditions in a power plant.
- Cost-effectiveness: The fin-tube heat exchanger design is cost-effective and widely available, making it a popular choice for ACCs.

In general, the use of fin-tube heat exchangers in ACCs provides an efficient and cost-effective solution for condensing the steam in power plants, while also providing the flexibility to adjust the cooling load based on the operating conditions and local climate.

Which metal is used to make a tube for an air-cooled condenser?

The tubes used in air-cooled condensers (ACCs) are typically made of copper or aluminum. Copper is a commonly used material for ACC tubes due to its high thermal conductivity and resistance to corrosion, which makes it ideal for use in high-temperature and harsh operating environments.

Aluminum is another popular material for ACC tubes because of its low cost, high strength, and good thermal conductivity. However, aluminum is more prone to corrosion than copper and may need to be coated or treated to improve its durability and longevity in the harsh operating environment of a power plant.

The choice between copper and aluminum for ACC tubes is often determined by factors such as cost, performance, and local availability of materials. In some cases, a combination of both materials may be used, with copper used for high-temperature areas and aluminum used for lower-temperature areas to optimize performance and cost.

What is the difference between an air-cooled condenser and a cooling tower?

- An air-cooled condenser (ACC) and a cooling tower are both used to remove heat from a power plant or industrial facility, but they operate in different ways and have some distinct differences.
- An air-cooled condenser is a type of heat exchanger that uses air to cool hot steam and condense it back into water. The steam is passed through tubes, typically made of copper or aluminum, which are surrounded by fins to increase the surface area for heat transfer. The fins and tubes are exposed to the surrounding air, which cools the steam and removes the heat from the system.
- A cooling tower, on the other hand, is a large structure used to cool hot water by exposing it to the air. Cooling towers work by drawing in hot water from an industrial process or power plant and spraying it into the air. As the water falls, it cools and is then collected and reused in the process or power plant. The cooling of the water is achieved through heat transfer from the water to the air.

In summary, an air-cooled condenser is used to cool steam and condense it back into water, while a cooling tower is used to cool hot water. Both systems serve the same purpose, but they have different methods of operation and are used in different parts of the power plant or industrial process.

How does dirt affect an air-cooled condenser?

Dirt on an air-cooled condenser can negatively affect its performance by reducing the efficiency of heat transfer. Dirt buildup on the fins of the condenser coils can insulate the coils, preventing air from flowing freely around them and reducing the rate at which heat is dissipated. Additionally, dirt and debris can clog the condenser's air intake, reducing the amount of air that can flow over the coils and further reducing heat dissipation. This can result in higher temperatures and reduced cooling capacity, potentially leading to equipment failure or reduced lifespan. Regular cleaning and maintenance of air-cooled condensers is important to prevent dirt buildup and maintain optimal performance.

What is an air-cooled condenser efficiency formula?

The efficiency of an air-cooled condenser can be calculated using the following formula:

Efficiency = (Qc / Wc) x 100%

Where:

- Qc = Cooling capacity (BTU/hr)
- Wc = Power consumption (watts)

This formula provides the efficiency of the air-cooled condenser as a percentage, indicating the amount of cooling capacity produced by a given amount of power consumed. The higher the efficiency, the more cooling capacity is produced per unit of power consumed. The efficiency of an air-cooled condenser can be influenced by various factors such as air flow, temperature, and dirt buildup, so it is important to regularly monitor and maintain the condenser to ensure optimal performance.

What is the acceptable vibration of air cooled condenser fans?

The acceptable vibration level for air-cooled condenser fans varies depending on the specific application and manufacturer specifications. In general, vibration levels for air-cooled condenser fans should be kept to a minimum to ensure smooth operation and prevent damage to the fan and surrounding equipment.

The American Society of Mechanical Engineers (ASME) standards provide guidelines for acceptable vibration levels, with typical values ranging from 0.1 to 1.0 inches per second (in/s) depending on the size and type of fan. It is important to consult the manufacturer's specifications for the specific air-cooled condenser fan in use to determine the acceptable vibration level.

Additionally, regular maintenance and monitoring of the fan and its components can help prevent excessive vibration and prolong the lifespan of the equipment.

What advantages are there to using a remote air-cooled condenser?

A remote air-cooled condenser has several advantages over a traditional, integrated air-cooled condenser:

- Flexibility: A remote air-cooled condenser can be located away from the main equipment, allowing for greater flexibility in terms of installation location and integration with other systems.
- Improved cooling performance: By separating the air-cooled condenser from the main equipment, the airflow around the condenser coils can be optimized for maximum cooling performance.
- Reduced noise: Remote air-cooled condensers can be located in a separate, quiet area away from the main equipment, reducing the overall noise level in the facility.
- Improved maintenance access: Remote air-cooled condensers are easier to access for maintenance and cleaning, which can improve their overall reliability and lifespan.
- Increased safety: By separating the air-cooled condenser from the main equipment, the risk of electrical shock or other safety hazards is reduced.
- Energy efficiency: By optimizing the airflow and placement of the condenser, energy consumption can be reduced, resulting in lower operating costs and improved sustainability.

Overall, a remote air-cooled condenser can provide improved performance, reduced noise, improved accessibility for maintenance, increased safety, and better energy efficiency compared to an integrated air-cooled condenser.

How can I stop a CEP pump after a turbine tripped in an air cool condenser?

You can stop a CEP (Condensate Extraction Pump) in an air-cooled condenser after a turbine trip by performing the following steps:

- Close the inlet and outlet valves of the pump to isolate it.
- Drain the pump casing to reduce the pressure inside.
- Switch off the electrical supply to the pump.
- Lock out the electrical switch to prevent accidental re-energization.

It's important to follow proper safety procedures and to make sure that the pump and the surrounding area are adequately secured before attempting to stop the pump.

What are the advantages of using a surface condenser over an air cooled condenser?

A surface condenser has several advantages over an air-cooled condenser:

- Increased Efficiency: Surface condensers are typically more efficient at condensing steam, leading to greater energy savings and increased plant efficiency.
- Reduced Water Usage: Surface condensers use cooling water, which is usually readily available, rather than air for cooling, which can result in reduced water usage and lower water treatment costs.
- Improved Reliability: Surface condensers are less susceptible to fouling and corrosion, which can improve reliability and reduce maintenance requirements compared to air-cooled condensers.

- Reduced Environmental Impact: Surface condensers are often more environmentally friendly than air-cooled condensers as they reduce the need for large fans and cooling towers, which can emit noise and pollutants.
- Space Requirements: Surface condensers typically require less space compared to air-cooled condensers, making them a good choice for installations where space is limited.

In summary, the choice between a surface condenser and an air-cooled condenser depends on the specific application and available resources, but surface condensers offer many advantages in terms of efficiency, reliability, and environmental impact.

How do I design an air-cooled condenser when I just know the incoming and desired outgoing temperature of the hot vapor fluid mixture?

Designing an air-cooled condenser requires the following steps:

- Calculate the heat load: Determine the amount of heat to be removed from the hot vapor fluid mixture by calculating the heat load, which is the difference between the incoming and desired outgoing temperatures of the fluid.
- Select the appropriate heat exchanger: Choose an air-cooled heat exchanger that is capable of handling the calculated heat load. Consider factors such as air flow rate, surface area, and fan power requirements.
- Determine the air flow rate: Calculate the required air flow rate to effectively cool the heat exchanger and remove the calculated heat load. Consider factors such as the air-side heat transfer coefficient, the air-side pressure drop, and the air temperature.
- Choose the fan and cooling tower: Select a fan and cooling tower that can provide the required air flow rate and ensure proper air distribution over the heat exchanger.
- Design the air ducts: Design the air ducts to ensure proper air flow distribution over the heat exchanger and minimize pressure drop.

- Consider any additional equipment: Consider any additional equipment that may be necessary, such as a cooling tower, pumps, and water treatment equipment.
- Perform thermal and hydraulic analysis: Perform thermal and hydraulic analysis to verify that the design will meet the desired outlet temperature and ensure adequate performance under various operating conditions.

It is recommended to consult a specialist or consult relevant standards for designing air-cooled condensers to ensure the design is appropriate and meets the necessary safety and performance requirements.

What is a multi-row tubing air-cooled condenser?

- A multi-row tubing air-cooled condenser typically consists of several rows of tubes arranged in parallel, with air passing over the tubes to cool the condensing steam. The tubes are connected to headers at each end and the headers are connected to the inlet and outlet piping. Air is drawn over the tubes by fans, and the cooled condensate is collected in a condensate return system and returned to the steam generator.
- In a multi-row tubing air-cooled condenser, multiple rows of tubes increase the overall heat transfer surface area, allowing for more effective cooling of the condensing steam. This design also provides increased flexibility and redundancy, as individual rows can be isolated for maintenance without affecting the overall performance of the condenser.
- The specific design of a multi-row tubing air-cooled condenser can vary depending on the application, with factors such as tube diameter, tube spacing, number of rows, and fan arrangement affecting the performance and efficiency of the condenser.

What are the advantages of using water cooled condensers over air cooled condensers in boilers?

There are several advantages to using water cooled condensers over air cooled condensers in boilers, including:

- Increased Efficiency: Water cooled condensers are able to transfer heat more efficiently than air cooled condensers, leading to higher thermal efficiency and improved performance of the boiler.
- Space-saving: Water cooled condensers take up less space compared to air cooled condensers, making them ideal for use in small or confined spaces.
- Lower Noise: Water cooled condensers are typically quieter than air cooled condensers, making them more suitable for use in noise-sensitive environments.
- Better Performance: Water cooled condensers can handle higher operating temperatures than air cooled condensers, allowing for more consistent and stable performance of the boiler.
- Lower Maintenance: Water cooled condensers have fewer moving parts compared to air cooled condensers, leading to lower maintenance requirements and reduced costs.
- Better Reliability: Water cooled condensers are generally more reliable than air cooled condensers, as they are less prone to failure due to exposure to the elements, such as dust and wind.

What are the disadvantages of a steam turbine generator set using water-cooled condensers as compared to those using air-cooled condensers?

There are several disadvantages of using water cooled condensers in steam turbine generator sets as compared to air cooled condensers, including:

- Increased Complexity: Water cooled condensers are more complex to design, install, and maintain compared to air cooled condensers, due to the need for a separate cooling water system and the potential for leaks and other issues.

- Higher Initial Cost: The initial cost of water cooled condensers is typically higher than that of air cooled condensers, due to the increased complexity and additional components required.
- Water Availability: The availability of a sufficient supply of cooling water is a critical requirement for water cooled condensers, whereas air cooled condensers do not require a separate water supply.
- Water Quality: The quality of the cooling water must be carefully monitored to prevent corrosion, scaling, and other issues that can impact the performance and longevity of the condenser.
- Water Disposal: The discharge of cooling water from water cooled condensers must be managed in an environmentally responsible manner, which can add additional costs and challenges.
- Maintenance: Water cooled condensers require more frequent maintenance compared to air cooled condensers, due to the need to monitor and maintain the cooling water system and the potential for leaks and other issues.

Why do we use cooling towers in thermal power stations, and what are its advantages over air-cooled condensers?

Cooling towers are used in thermal power stations to cool the hot water used in the condenser, after it has been used to condense steam from the turbine. Cooling towers offer several advantages over air-cooled condensers, including:

- Improved Efficiency: Cooling towers allow for the continuous use of cooling water, reducing the energy consumption and increasing the overall efficiency of the thermal power station.
- Space-saving: Cooling towers take up less space compared to air-cooled condensers, as they can be located away from the main power plant structure.
- Water Savings: Cooling towers can conserve water compared to air-cooled condensers, as only a portion of the water used for

cooling is lost to evaporation, while the rest can be reused.

- Lower Operating Costs: Cooling towers generally have lower operating costs compared to air-cooled condensers, as they do not require the use of fans or other mechanical components.
- Improved Reliability: Cooling towers are generally more reliable than air-cooled condensers, as they have fewer moving parts and are less prone to failure due to exposure to the elements, such as wind and dust.
- Environmentally Friendly: Cooling towers can be designed and operated in an environmentally responsible manner, reducing the impact on the environment compared to air-cooled condensers, which can release hot air into the atmosphere.

Can an air cooled condenser be used in OTEC to cool ammonia (to liquid) instead of pumping cool water from deeper ocean?

Yes, an air cooled condenser can be used in an Ocean Thermal Energy Conversion (OTEC) system to cool ammonia, instead of pumping cool water from the deeper ocean. However, it is important to note that the efficiency of an air cooled condenser in an OTEC system will likely be lower compared to a water cooled condenser, as air cooling is generally less efficient than water cooling. Additionally, air cooled condensers are also more susceptible to environmental conditions such as temperature, humidity, and air flow, which can impact their performance. Therefore, careful consideration of the environmental conditions and cooling requirements must be taken into account when selecting an air cooled condenser for use in an OTEC system.

Can condensate from an air-cooled condenser be fed directly into the boiler?

Yes, the condensate from an air-cooled condenser can be fed directly into the boiler, as long as it meets the required quality standards for boiler feed water. The quality of the condensate is influenced by several factors such as temperature, pressure, and contamination, and it must be treated to remove impurities before it can be used in the boiler. In some cases, additional pre-treatment or conditioning of the condensate may be required to ensure that it is suitable for use as boiler feed water. This may include processes such as filtration, deaeration, and chemical treatment. It is recommended to consult with a qualified engineer to determine the specific requirements for using condensate from an air-cooled condenser in a particular boiler application.

How do I maintain an air cooled condenser?

To maintain an air cooled condenser, it is important to keep the area around the condenser clean and clear of debris, as this will allow the air to flow freely over the tubes or coils and improve the condenser's efficiency. The fan and motors should also be checked periodically to ensure they are in good working order. Additionally, the tubes or coils may need to be cleaned if they become clogged with dirt or other debris.

How does an air-cooled condenser transfer heat from the refrigerant to the atmosphere?

A: An air-cooled condenser transfers heat from the refrigerant to the atmosphere by circulating air over the finned tubes, where the refrigerant releases heat as it condenses. The air carries the heat away from the refrigerant and expels it to the atmosphere.

Q: What is the role of the fans in an air-cooled condenser?

A: The fans serve to circulate the air over the finned tubes, which helps transfer the heat from the refrigerant to the atmosphere.

Q: What are some of the latest technologies in air-cooled condenser systems?

A: Some of the latest technologies in air-cooled condenser systems include variable speed fans, electronic expansion valves, and microchannel condenser coils.

Q: How do these technologies improve the efficiency of air-cooled condenser systems?

A: These technologies improve the efficiency of air-cooled condenser systems by providing better control over the refrigerant flow and heat transfer. For example, variable speed fans can adjust their speed based on the heat load, reducing energy consumption and improving efficiency.

Q: What are the industry standards for air-cooled condensers?

A: The industry standards for air-cooled condensers are typically set by organizations such as AHRI (Air-Conditioning, Heating, and Refrigeration Institute) and ASHRAE (American Society of Heating, Refrigerating, and Air-Conditioning Engineers).

Q: Why are industry standards important for air-cooled condensers?

A: Industry standards are important for air-cooled condensers because they provide guidelines for design, performance, and efficiency, ensuring that air-cooled condensers are safe, reliable, and energy-efficient. Adherence to industry standards also promotes consistency and compatibility across different types of air-cooled condenser systems.

How do I troubleshoot an air cooled condenser?

If an air cooled condenser is not working properly, there are a few common issues that may be causing the problem. These include: a malfunctioning fan or motor, a clogged or damaged condenser coil, or a refrigerant leak. To troubleshoot the issue, you can check the fan and motor to ensure they are functioning correctly, clean or repair the condenser coil as needed, and check for any leaks in the refrigerant system.

What are the different types of air cooled condensers?

There are several types of air cooled condensers, including:

- Finned tube condensers: These condensers use a series of tubes with fins attached to increase the surface area for heat transfer.
- Plate fin condensers: These condensers consist of a series of flat plates with fins attached to them, which are stacked together to form a compact, efficient heat exchanger.
- Tube and fin condensers: These condensers use a combination of tubes and fins to transfer heat from the fluid to the air.
- Microchannel condensers: These condensers use a series of small channels or tubes to transfer heat from the fluid to the air, and are often used in compact refrigeration systems.

How do I choose the right air cooled condenser for my application?

There are a few factors to consider when choosing an air cooled condenser for your application. These include:

- The size and capacity of the condenser: Make sure to choose a condenser that is large enough to handle the heat load of your system.
- The operating conditions: Consider the ambient temperature and humidity in the area where the condenser will be used, as well as any other factors that may affect its performance, such as the presence of dust or other contaminants.
- The type of refrigerant: Make sure to choose a condenser that is compatible with the type of refrigerant you will be using.
- The efficiency of the condenser: Look for a condenser that is designed to be as efficient as possible, as this can help to reduce energy costs and improve the overall performance of your system.

What are the common applications of air cooled condensers?

Air cooled condensers are used in a wide range of applications, including:

- Refrigeration and air conditioning systems
- Industrial processes
- Power plants
- Chemical plants
- Food and beverage processing
- HVAC systems

Do you have any other specific questions about air cooled condensers that I can help with?

How do air cooled condensers compare to water cooled condensers?

Air cooled condensers and water cooled condensers are both used to transfer heat from a refrigerant or other fluid to the air or water, respectively. Water cooled condensers are generally more efficient at removing heat, as water has a higher heat capacity than air and is therefore better able to absorb and transfer heat. However, air cooled condensers are typically simpler and less expensive to install and maintain, as they do not require a water source or water treatment system.

What are the main components of an air-cooled condenser?

The main components of an air-cooled condenser are:

- Heat Exchanger: The heat exchanger is the core component of the air-cooled condenser, where the refrigerant is cooled by passing over the fins or tubes that are exposed to the air flow.
- Fans: Fans are used to move air over the heat exchanger and facilitate heat transfer from the refrigerant to the air.
- Motor: A motor drives the fans to circulate air through the condenser.
- Fan Guard: A fan guard is used to protect the fans and prevent debris from entering the condenser.
- Air Ducts: Air ducts are used to direct the air flow over the heat exchanger and provide proper ventilation.
- Filter: A filter may be used to remove particles from the air flow and improve the performance of the condenser.
- Control System: A control system is used to regulate the operation of the fans and monitor the performance of the air-cooled condenser.
- Refrigerant Connections: Refrigerant connections are used to connect the air-cooled condenser to the refrigerant system.

These are the main components of an air-cooled condenser, and some models may include additional components or features, depending on the specific requirements of the application.

What is the fouling factor for an air-cooled condenser?

- The fouling factor for an air-cooled condenser refers to the decrease in heat transfer efficiency caused by the accumulation of dirt, debris, and other materials on the surface of the heat exchanger. The fouling factor is expressed as a percentage, and it represents the reduction in heat transfer efficiency compared to a clean surface.
- The fouling factor for an air-cooled condenser can vary widely, depending on the environment and the type of application. Factors that can influence the fouling factor include the location of the condenser, the temperature and humidity of the air, the presence of pollutants or other contaminants, and the level of maintenance and cleaning performed on the condenser.
- It is important to monitor the fouling factor for an air-cooled condenser, as a high fouling factor can lead to decreased efficiency, increased energy costs, and reduced performance of the condenser. To minimize the fouling factor, regular cleaning and maintenance should be performed on the condenser, and measures should be taken to minimize exposure to contaminants.

What are the advantages and disadvantages of using a shell and tube type condenser over an air-cooled or water cooled condenser?

Advantages of Shell and Tube Condensers:

- High Heat Transfer Efficiency: Shell and tube condensers have a high heat transfer efficiency, making them well-suited for applications where high heat transfer rates are required.
- Durability: The shell and tube design is rugged and durable, and the tubes can be easily replaced if necessary.
- Versatility: Shell and tube condensers can be used with a variety of fluids and refrigerants, making them versatile for a wide range of applications.
- Scalability: Shell and tube condensers can be designed and manufactured in a variety of sizes and capacities, making them well-suited for large-scale applications.

- Scalability: Shell and tube condensers can be designed and manufactured in a variety of sizes and capacities, making them well-suited for large-scale applications.

Disadvantages of Shell and Tube Condensers:

- Complexity: The shell and tube design can be complex and difficult to maintain, and repairs may require specialized equipment and expertise.
- Cost: Shell and tube condensers can be more expensive to manufacture and install than air-cooled or water-cooled condensers, making them less cost-effective for smaller applications.
- Maintenance: Shell and tube condensers require regular maintenance, including cleaning and monitoring of the tubes, which can be time-consuming and costly.

In summary, shell and tube condensers offer high heat transfer efficiency and versatility, but they are more complex and expensive than air-cooled or water-cooled condensers. The choice between a shell and tube condenser, an air-cooled condenser, or a water-cooled condenser will depend on the specific requirements of the application and the trade-offs between efficiency, cost, and ease of maintenance.

What are the advantages and disadvantages of tubular and finned air-cooled condensers in nuclear power plant applications?

Advantages of Tubular Air-Cooled Condensers:

- Compact Design: Tubular air-cooled condensers are compact in design and take up less space compared to other types of air-cooled condensers, making them ideal for use in limited-space applications like nuclear power plants.
- High heat transfer efficiency: Tubular air-cooled condensers have high heat transfer efficiency due to their large surface area and close proximity of tubes to the air flow.

- Reliable: Tubular air-cooled condensers are known for their reliability and durability, making them ideal for use in demanding applications like nuclear power plants where continuous operation is critical.

Advantages of Finned Air-Cooled Condensers:

- High heat dissipation: Finned air-cooled condensers have high heat dissipation capacity due to the large surface area provided by the fins.
- Increased air flow: The finned design increases the air flow over the surface of the tubes, resulting in improved heat transfer.
- Easy maintenance: Finned air-cooled condensers are relatively simple in design and easy to maintain, making them an ideal choice for use in nuclear power plants where regular maintenance and upkeep are required.

Disadvantages of Tubular Air-Cooled Condensers:

- High initial cost: The compact design and high heat transfer efficiency of tubular air-cooled condensers come at a higher initial cost compared to other types of air-cooled condensers.
- More complex design: The design of tubular air-cooled condensers is more complex compared to other types of air-cooled condensers, making them more challenging to manufacture and maintain.

Disadvantages of Finned Air-Cooled Condensers:

- Low heat transfer efficiency: The finned design can result in lower heat transfer efficiency compared to other types of air-cooled condensers, especially when air flow is restricted.
- Increased maintenance: The finned design can result in increased maintenance requirements, particularly in terms of cleaning the fins and ensuring proper air flow.

In conclusion, the choice between tubular and finned air-cooled condensers will depend on the specific requirements of the nuclear power plant and the trade-off between initial cost, heat transfer efficiency, reliability, and ease of maintenance.

How can I calculate the pressure drop across the bundle of a single row air cooled steam condenser?

- The pressure drop across the bundle of a single row air cooled steam condenser can be calculated using the following formula:

$\Delta P = \Delta Pf + \Delta Pc$

where ΔPf is the pressure drop due to friction and ΔPc is the pressure drop due to air cooling.

- The pressure drop due to friction can be calculated using the Darcy-Weisbach equation:

$\Delta Pf = f * (L/D) * v^2 / (2 * g)$

where f is the friction factor, L is the length of the bundle, D is the diameter of the tubes, v is the velocity of the air, and g is the acceleration due to gravity.

- The pressure drop due to air cooling can be calculated using the formula:

$\Delta Pc = \rho * c * (T1 - T2) / T1$

where ρ is the density of the air, c is the specific heat of the air, T1 is the initial temperature of the air, and T2 is the final temperature of the air after passing through the bundle.

It is important to note that the pressure drop across the bundle can be affected by a number of factors, including the size and configuration of the tubes, the air velocity, the air temperature, and the air humidity. In practice, it may be necessary to experimentally determine the pressure drop across the bundle in order to obtain accurate results.

What is the solution to the problem of an air-cooled condenser not having any wind in front of it and not being able to install any kind of fan or blower?

If an air-cooled condenser is not getting enough air flow and it is not possible to install a fan or blower, there are a few potential solutions:

- Relocate the condenser: If possible, the air-cooled condenser can be relocated to an area with better air flow. This may require some modifications to the refrigerant piping, but it can be an effective solution to the problem.
- Install passive air flow devices: Passive air flow devices, such as louvers or deflectors, can be installed to redirect air flow over the condenser. These devices do not require a fan or blower and can be a low-cost solution to the problem.
- Use a water-cooled condenser: If the air flow problem cannot be resolved, a water-cooled condenser may be an alternative solution. Water-cooled condensers do not rely on air flow and can be a viable option in areas with limited air flow.
- Implement process changes: If possible, process changes can be made to reduce the heat load on the condenser and reduce the demand for cooling. This can reduce the need for air flow over the condenser and may be a solution in certain applications.

These are some potential solutions to the problem of an air-cooled condenser not having enough air flow, but the best solution will depend on the specific requirements of the application and the available resources.

What are the advantages and disadvantages of air cooled and water cooled condensers for distillation?

Advantages of Air-Cooled Condensers for Distillation:

- Cost-effective: Air-cooled condensers are often less expensive to install and maintain than water-cooled condensers, making them a cost-effective option for some distillation applications.

- Easy to maintain: Air-cooled condensers do not require a water supply or water treatment system, making them easier to maintain than water-cooled condensers.
- Versatile: Air-cooled condensers can be used in a wide range of distillation applications and can be designed to meet specific heat transfer requirements.

Disadvantages of Air-Cooled Condensers for Distillation:

- Limited heat transfer capacity: Air-cooled condensers have a limited heat transfer capacity and may not be suitable for high heat loads in distillation applications.
- Sensitive to ambient conditions: Air-cooled condensers can be sensitive to ambient conditions, such as temperature and humidity, and may not perform optimally in extreme conditions.

Advantages of Water-Cooled Condensers for Distillation:

- High heat transfer capacity: Water-cooled condensers have a high heat transfer capacity and are well-suited for high heat loads in distillation applications.
- Reliable performance: Water-cooled condensers are less sensitive to ambient conditions than air-cooled condensers and can provide more reliable performance in extreme conditions.
- Energy-efficient: Water-cooled condensers can be more energy-efficient than air-cooled condensers in some distillation applications, as they can use waste heat from the distillation process to preheat the cooling water.

Disadvantages of Water-Cooled Condensers for Distillation:

- More expensive: Water-cooled condensers can be more expensive to install and maintain than air-cooled condensers, making them a less cost-effective option for some distillation applications.

- Complex to maintain: Water-cooled condensers require a water supply and water treatment system, making them more complex to maintain than air-cooled condensers.

In conclusion, the choice between an air-cooled or water-cooled condenser for distillation will depend on the specific requirements of the application and the trade-offs between cost, performance, and maintenance.

What are the advantages of using an induction motor in an air cooled condenser over other types of AC motors?

Advantages of using an Induction Motor in an Air-Cooled Condenser:

- Reliability: Induction motors are known for their reliability and durability, making them an ideal choice for use in demanding industrial applications, such as air-cooled condensers.
- Efficiency: Induction motors are highly efficient and can operate at high levels of power output for extended periods of time, making them well-suited for use in air-cooled condensers.
- Cost-effective: Induction motors are relatively inexpensive compared to other types of AC motors and are widely available, making them a cost-effective choice for use in air-cooled condensers.
- Easy to maintain: Induction motors are simple in design and easy to maintain, making them an ideal choice for use in air-cooled condensers that require regular maintenance and upkeep.
- Versatile: Induction motors can be designed for use in a wide range of applications, including air-cooled condensers, and can be configured to meet specific performance requirements.

These advantages make induction motors a popular choice for use in air-cooled condensers and other industrial applications where reliability, efficiency, and cost-effectiveness are important considerations.

Q: What is a refrigeration air-cooled condenser?

A: A refrigeration air-cooled condenser is a type of condenser used in refrigeration systems to reject heat from the refrigerant and transfer it to the atmosphere.

Q: How does a refrigeration air-cooled condenser work?

A: A refrigeration air-cooled condenser consists of a series of finned tubes through which the refrigerant flows and fans that circulate air over the tubes to dissipate heat. The refrigerant releases heat as it condenses and the air-cooled condenser expels this heat to the atmosphere.

Q: What are the components of an air-cooled condenser?

A: The components of an air-cooled condenser include the finned tubes, fans, motor, and control system.

Q: What is the role of the finned tubes in an air-cooled condenser?

A: The finned tubes serve as the surface over which the refrigerant releases heat as it condenses. The fins on the tubes increase the surface area and help transfer heat to the air.

Can air cooled condensers be used in outdoor applications?

Yes, air cooled condensers can be used in outdoor applications, provided that the condenser is protected from the elements and the ambient temperature is within the operating range of the condenser. It is important to choose a condenser that is designed to operate in outdoor environments and is capable of withstanding the conditions it will be exposed to.

What are the benefits of using an air cooled condenser in a refrigeration system?

There are several benefits to using an air cooled condenser in a refrigeration system, including:

- Simplicity: Air cooled condensers are relatively simple and easy to install, operate, and maintain.
- Cost: Air cooled condensers are generally less expensive to install and maintain than water cooled condensers.
- Compact size: Air cooled condensers are generally smaller and more compact than water cooled condensers, making them a

good choice for applications where space is limited.

- Flexibility: Air cooled condensers can be used in a wide range of applications and environments, and do not require a water source.

How do I calculate the size of an air cooled condenser?

To calculate the size of an air cooled condenser, you will need to determine the heat load of the system and the desired temperature of the refrigerant as it exits the condenser. You can then use a formula or a condenser sizing tool to determine the required size of the condenser. Some factors to consider when sizing an air cooled condenser include:

- The type and amount of refrigerant
- The ambient temperature and humidity
- The desired temperature of the refrigerant
- The type and size of the evaporator
- The capacity of the compressor

How do I install an air cooled condenser?

Installing an air cooled condenser typically involves the following steps:

- Choose a location for the condenser: The condenser should be located in a well-ventilated area where it will be protected from the elements.
- Prepare the site: Clear the area around the condenser and level the ground to ensure that the condenser is stable and secure.
- Install the condenser: Follow the manufacturer's instructions to install the condenser, including attaching the mounting brackets and connecting the refrigerant lines.
- Connect the electrical wiring: Connect the condenser to a power source and wire the fan and motor according to the manufacturer's instructions.

- Test the system: Once the condenser is installed, test the system to ensure that it is operating correctly.

How do I know if my air cooled condenser is working properly?

There are a few signs that can indicate that an air cooled condenser is working properly:

- The condenser is running: If the fan and motor are functioning correctly, the condenser should be running and blowing air over the tubes or coils.
- The refrigerant temperature is within range: The temperature of the refrigerant as it exits the condenser should be within the specified range for the system.
- The system is operating efficiently: If the system is operating efficiently, the compressor should be running smoothly and the system should be able to maintain the desired temperature.

If you notice any issues with the operation of the condenser or the system as a whole, it may be necessary to troubleshoot and repair the problem.

What are the common causes of problems with air cooled condensers?

There are several common causes of problems with air cooled condensers, including:

- Refrigerant leaks: If the refrigerant system is leaking, it can cause the condenser to lose efficiency and may even cause the system to fail completely.
- Clogged or damaged condenser coils: If the condenser coils become clogged or damaged, it can affect the performance of the condenser and may even cause it to fail.
- Malfunctioning fan or motor: If the fan or motor is not functioning properly, it can cause the condenser to fail or operate inefficiently.

- Incorrect refrigerant charge: If the refrigerant charge in the system is too low or too high, it can cause the condenser to operate inefficiently or even fail.

How do I clean an air cooled condenser?

Cleaning an air cooled condenser typically involves the following steps:

- Turn off the power: Before cleaning the condenser, make sure to turn off the power to the unit to prevent electrical accidents.
- Remove any debris: Use a broom or brush to remove any debris, such as leaves or dirt, from the area around the condenser.
- Clean the condenser coils: Use a soft brush or cloth to gently remove any dirt or debris from the condenser coils. You can also use a coil cleaning solution or water to help remove any stubborn dirt or grime.
- Clean the fan blades: Use a cloth or brush to remove any dirt or debris from the fan blades.
- Reassemble the condenser: Once the condenser has been cleaned, reassemble the unit and turn the power back on.

It is important to be careful when cleaning an air cooled condenser, as the tubes or coils can be delicate and may be damaged if handled roughly.

How often should an air cooled condenser be cleaned?

The frequency with which an air cooled condenser should be cleaned will depend on the environment in which it is used. In general, it is a good idea to clean the condenser at least once a year, or more often if it is located in an area with high levels of dust or other contaminants. Regular cleaning can help to improve the efficiency of the condenser and extend its lifespan.

Can an air cooled condenser be repaired if it is damaged?

In some cases, an air cooled condenser can be repaired if it is damaged. However, the extent of the damage and the cost of the repair will vary depending on the specific problem. If the damage is minor, it may be possible to repair the condenser by cleaning or replacing damaged components. However, if the damage is more severe, it may be more cost-effective to replace the condenser rather than trying to repair it.

How long does an air cooled condenser last?

The lifespan of an air cooled condenser will depend on a variety of factors, including the quality of the unit, the operating conditions, and the level of maintenance it receives. In general, air cooled condensers are designed to last for many years, and with proper care and maintenance, they can have a lifespan of 10 years or more.

Are there any safety concerns to be aware of when working with an air cooled condenser?

Yes, there are a few safety concerns to be aware of when working with an air cooled condenser:

- Electrical safety: When working with the electrical components of an air cooled condenser, it is important to follow proper safety procedures to prevent electrical accidents.
- Refrigerant safety: When handling refrigerant, it is important to wear protective gear and follow proper safety procedures to avoid inhaling the refrigerant or coming into contact with it.
- Handling heavy equipment: Air cooled condensers can be heavy and may require special equipment to lift or move. Make sure to follow proper safety procedures and use the appropriate equipment when handling the condenser.

Can an air cooled condenser be used in a heat pump system?

Yes, an air cooled condenser can be used in a heat pump system. A heat pump is a type of system that uses a refrigeration cycle to transfer heat from one location to another, either from the air or ground to the interior of a building, or vice versa. An air cooled condenser is used to transfer heat from the refrigerant to the air, and can be used in either the heating or cooling mode, depending on the desired direction of heat transfer.

Can an air cooled condenser be used in a geothermal system?

An air cooled condenser can be used in a geothermal system, but it is generally not the most efficient choice. Geothermal systems use the constant temperature of the earth as a heat source or sink, and typically use a water cooled condenser to transfer heat to or from the refrigerant. An air cooled condenser could be used in a geothermal system, but it may not be as efficient as a water cooled condenser, and may require a larger size to handle the heat load.

Are there any alternative cooling methods to air cooled condensers?

Some alternatives to air cooled condensers include water cooled condensers, evaporative cooling, and ice storage systems. Water cooled condensers use water to transfer heat from the refrigerant to the air, and are generally more efficient than air cooled condensers. Evaporative cooling systems use the evaporation of water to cool the air, and are often used in dry, hot environments. Ice storage systems use ice to store cooling energy, which can be used to cool a building or process during periods of high demand.

What are the benefits of using an air cooled condenser in an industrial process?

There are several benefits to using an air cooled condenser in an industrial process, including:

- Simplicity: Air cooled condensers are relatively simple and easy to install, operate, and maintain.
- Cost: Air cooled condensers are generally less expensive to install and maintain than water cooled condensers.
- Compact size: Air cooled condensers are generally smaller and more compact than water cooled condensers, making them a good choice for applications where space is limited.
- Flexibility: Air cooled condensers can be used in a wide range of applications and environments, and do not require a water source.
- Customization: Air cooled condensers can be customized to meet the specific needs of the industrial process, including the

type and amount of refrigerant, the operating temperature range, and the required capacity.

What are the benefits of using an air cooled condenser in a power plant?

There are several benefits to using an air cooled condenser in a power plant, including:

- Simplicity: Air cooled condensers are relatively simple and easy to install, operate, and maintain.
- Cost: Air cooled condensers are generally less expensive to install and maintain than water cooled condensers.
- Compact size: Air cooled condensers are generally smaller and more compact than water cooled condensers, making them a good choice for applications where space is limited.
- Flexibility: Air cooled condensers can be used in a wide range of power plant applications, and do not require a water source.
- Customization: Air cooled condensers can be customized to meet the specific needs of the power plant, including the type and amount of refrigerant, the operating temperature range, and the required capacity.

Can an air cooled condenser be used in a chemical plant?

Yes, an air cooled condenser can be used in a chemical plant. Air cooled condensers are often used in chemical plants to cool refrigerants or other fluids used in the manufacturing process, or to provide cooling for process equipment. Air cooled condensers are generally simple and easy to install, operate, and maintain, and can be customized to meet the specific needs of the chemical plant.

Can an air cooled condenser be used in a food and beverage processing facility?

Yes, an air cooled condenser can be used in a food and beverage processing facility. Air cooled condensers are often used in food and beverage processing facilities to provide cooling for refrigeration equipment, such as refrigerators and freezers, or to cool process fluids used in the production process. Air cooled condensers are generally simple and easy to install, operate, and maintain, and can be customized to meet the specific needs of the food and beverage processing facility.

Can an air cooled condenser be used in an HVAC system?

Yes, an air cooled condenser can be used in an HVAC system. HVAC stands for heating, ventilation, and air conditioning, and refers to systems that are used to control the temperature, humidity, and air quality in a building or other enclosed space. An air cooled condenser is used to transfer heat from the refrigerant to the air, and can be used in either the heating or cooling mode, depending on the desired direction of heat transfer.

Are there any limitations to using an air cooled condenser?

There are a few limitations to using an air cooled condenser:

- Operating temperature range: Air cooled condensers are generally limited to a certain operating temperature range, and may not be suitable for use in extremely cold or hot environments.
- Ambient temperature and humidity: The performance of an air cooled condenser can be affected by the ambient temperature and humidity, and it may not be as efficient in humid environments.
- Water availability: Air cooled condensers do not require a water source, but water cooled condensers may be a more efficient choice in applications where a water source is readily available.
- Size and capacity: Air cooled condensers are generally limited in size and capacity compared to water cooled condensers, and may not be suitable for large or high heat load

What is the difference between an air cooled condenser and an air cooled heat exchanger?

- An air cooled condenser and an air cooled heat exchanger are similar in that they both use air to transfer heat from one fluid to another, but they have different functions and are used in different types of systems.

- An air cooled condenser is used to transfer heat from a refrigerant or other fluid to the air, and is commonly used in refrigeration and air conditioning systems to reject heat from the system.
- An air cooled condenser is used to transfer heat from a refrigerant or other fluid to the air, and is commonly used in refrigeration and air conditioning systems to reject heat from the system.
- An air cooled heat exchanger, on the other hand, is used to transfer heat from one fluid to another without mixing the fluids, and is commonly used in industrial processes to transfer heat between process fluids or to provide cooling for process equipment.

How does an air cooled condenser work?

An air cooled condenser works by using a fan to blow air over tubes or coils that contain the refrigerant or other fluid to be cooled. As the air passes over the tubes or coils, it absorbs heat from the fluid, and the cooled fluid is then returned to the system. The heat absorbed by the air is dissipated into the surrounding environment, and the cooled air is then re-circulated back over the tubes or coils.

What are the different types of air cooled condensers?

There are several different types of air cooled condensers, including:

- Tube and fin condensers: These condensers consist of tubes with fins attached to them to increase the surface area for heat transfer.
- Plate fin condensers: These condensers consist of flat plates with fins attached to them to increase the surface area for heat transfer.

- Spiral condensers: These condensers consist of spiral-shaped tubes with fins attached to them to increase the surface area for heat transfer.

How is an air cooled condenser sized?

An air cooled condenser is sized based on the heat load of the system and the desired temperature of the refrigerant as it exits the condenser. The size of the condenser is determined using a formula or a condenser sizing tool, which takes into account factors such as the type and amount of refrigerant, the ambient temperature and humidity, the desired temperature of the refrigerant, the type and size of the evaporator, and the capacity of the compressor.

How is an air cooled condenser maintained?

An air cooled condenser should be maintained regularly to ensure that it is operating efficiently and to extend its lifespan. Some common maintenance tasks for an air cooled condenser include:

- Cleaning: The condenser should be cleaned regularly to remove dirt and debris that can build up on the tubes or coils and affect the performance of the unit.
- Checking and tightening connections: The connections between the tubes or coils and the rest of the system should be checked and tightened as needed to ensure that the system is sealed and there are no leaks.
- Checking and replacing the fan belt: The fan belt should be checked regularly for wear and replaced if it becomes damaged or worn.

How is an air cooled condenser installed?

Installing an air cooled condenser typically involves the following steps:

- Choose a location for the condenser: The condenser should be located in a well-ventilated area where it will be protected from the elements.

- Prepare the site: Clear the area around the condenser and level the ground to ensure that the condenser is stable and secure.
- Install the condenser: Follow the manufacturer's instructions to install the condenser, including attaching the mounting brackets and connecting the refrigerant lines.
- Connect the electrical wiring: Connect the condenser to a power source and wire the fan and motor according to the manufacturer's instructions.
- Test the system: Once the condenser is installed, test the system to ensure that it is operating correctly.

It is important to follow the manufacturer's instructions and adhere to any local building codes when installing an air cooled condenser.

How is an air cooled condenser repaired?

If an air cooled condenser is not functioning properly, it may need to be repaired. The specific steps for repairing an air cooled condenser will depend on the problem, but may include:

- Troubleshooting the issue: The first step in repairing an air cooled condenser is to determine the cause of the problem. This may involve checking for leaks, inspecting the condenser coils, or testing the electrical components.
- Cleaning or replacing damaged components: If the condenser coils are clogged or damaged, they may need to be cleaned or replaced. Similarly, if the fan or motor is not functioning properly, it may need to be repaired or replaced.

What are the advantages of using an air cooled condenser in a refrigeration system?

There are several advantages to using an air cooled condenser in a refrigeration system, including:

- Simplicity: Air cooled condensers are relatively simple and easy to install, operate, and maintain.
- Cost: Air cooled condensers are generally less expensive to install and maintain than water cooled condensers.
- Compact size: Air cooled condensers are generally smaller and more compact than water cooled condensers, making them a good choice for applications where space is limited.
- Flexibility: Air cooled condensers can be used in a wide range of refrigeration applications and environments, and do not require a water source.
- Customization: Air cooled condensers can be customized to meet the specific needs of the refrigeration system, including the type and amount of refrigerant, the operating temperature range, and the required capacity.

How does an air cooled condenser compare to a water cooled condenser in terms of efficiency?

Air cooled condensers are generally less efficient than water cooled condensers because the air is not as effective at transferring heat as water. However, the efficiency of an air cooled condenser can be improved by increasing the surface area of the tubes or coils, using a high-efficiency fan, or using a microchannel condenser. In addition, the efficiency of an air cooled condenser can be affected by the ambient temperature and humidity, and it may not be as efficient in humid environments.

How does an air cooled condenser compare to an evaporative cooled condenser in terms of efficiency?

An evaporative cooled condenser is generally more efficient than an air cooled condenser because it uses the evaporation of water to cool the air, which is a more effective way of transferring heat than using dry air.

However, evaporative cooled condensers are typically only used in dry, hot environments, and may not be suitable for use in humid environments. In addition, evaporative cooled condensers require a water source, and may not be practical in some locations.

How does an air cooled condenser compare to an ice storage system in terms of efficiency?

An ice storage system is generally more efficient than an air cooled condenser because it uses ice to store cooling energy, which can be used to cool a building or process during periods of high demand. However, ice storage systems require a large amount of space to store the ice, and may not be practical in some applications. In addition, ice storage systems are typically only used in applications with a high cooling load, and may not be suitable for smaller or lower heat load applications.

What are the advantages of using an air cooled condenser in a power plant?

There are several advantages to using an air cooled condenser in a power plant, including:

- Simplicity: Air cooled condensers are relatively simple and easy to install, operate, and maintain.
- Cost: Air cooled condensers are generally less expensive to install and maintain than water cooled condensers.
- Compact size: Air cooled condensers are generally smaller and more compact than water cooled condensers, making them a good choice for applications where space is limited.
- Flexibility: Air cooled condensers can be used in a wide range of power plant applications, and do not require a water source.
- Customization: Air cooled condensers can be customized to meet the specific needs of the power plant, including the type and amount of refrigerant, the operating temperature range, and the required capacity.

What are the advantages of using an air cooled condenser in a chemical plant?

There are several advantages to using an air cooled condenser in a chemical plant, including:

- Simplicity: Air cooled condensers are relatively simple and easy to install, operate, and maintain.
- Cost: Air cooled condensers are generally less expensive to install and maintain than water cooled condensers.
- Compact size: Air cooled condensers are generally smaller and more compact than water cooled condensers, making them a good choice for applications where space is limited.

What are the advantages of using an air cooled condenser in a food and beverage processing facility?

There are several advantages to using an air cooled condenser in a food and beverage processing facility, including:

- Simplicity: Air cooled condensers are relatively simple and easy to install, operate, and maintain.
- Cost: Air cooled condensers are generally less expensive to install and maintain than water cooled condensers.
- Compact size: Air cooled condensers are generally smaller and more compact than water cooled condensers, making them a good choice for applications where space is limited.
- Flexibility: Air cooled condensers can be used in a wide range of food and beverage processing applications, and do not require a water source.
- Customization: Air cooled condensers can be customized to meet the specific needs of the food and beverage processing facility, including the type and amount of refrigerant, the operating temperature range, and the required capacity.

What are the advantages of using an air cooled condenser in an HVAC system?

There are several advantages to using an air cooled condenser in an HVAC system, including:

- Simplicity: Air cooled condensers are relatively simple and easy to install, operate, and maintain.

- Cost: Air cooled condensers are generally less expensive to install and maintain than water cooled condensers.
- Compact size: Air cooled condensers are generally smaller and more compact than water cooled condensers, making them a good choice for applications where space is limited.

What are the limitations of using an air cooled condenser in a refrigeration system?

There are a few limitations to using an air cooled condenser in a refrigeration system:

- Operating temperature range: Air cooled condensers are generally limited to a certain operating temperature range, and may not be suitable for use in extremely cold or hot environments.
- Ambient temperature and humidity: The performance of an air cooled condenser can be affected by the ambient temperature and humidity, and it may not be as efficient in humid environments.
- Water availability: Air cooled condensers do not require a water source, but water cooled condensers may be a more efficient choice in applications where a water source is readily available.
- Size and capacity: Air cooled condensers are generally limited in size and capacity compared to water cooled condensers, and may not be suitable for large or high heat load applications.

What are the limitations of using an air cooled condenser in an air conditioning system?

There are a few limitations to using an air cooled condenser in an air conditioning system:

- Operating temperature range: Air cooled condensers are generally limited to a certain operating temperature range, and may not be suitable for use in extremely cold or hot environments.

- Ambient temperature and humidity: The performance of an air cooled condenser can be affected by the ambient temperature and humidity, and it may not be as efficient in humid environments.
- Water availability: Air cooled condensers do not require a water source, but water cooled condensers may be a more efficient choice in applications where a water source is readily available.

What are the limitations of using an air cooled condenser in a power plant?

There are a few limitations to using an air cooled condenser in a power plant:

- Operating temperature range: Air cooled condensers are generally limited to a certain operating temperature range, and may not be suitable for use in extremely cold or hot environments.
- Ambient temperature and humidity: The performance of an air cooled condenser can be affected by the ambient temperature and humidity, and it may not be as efficient in humid environments.
- Water availability: Air cooled condensers do not require a water source, but water cooled condensers may be a more efficient choice in applications where a water source is readily available.
- Size and capacity: Air cooled condensers are generally limited in size and capacity compared to water cooled condensers, and may not be suitable for large or high heat load applications.

What are the limitations of using an air cooled condenser in a chemical plant?

There are a few limitations to using an air cooled condenser in a chemical plant:

- Operating temperature range: Air cooled condensers are generally limited to a certain operating temperature range, and may not be suitable for use in extremely cold or hot environments.

- Ambient temperature and humidity: The performance of an air cooled condenser can be affected by the ambient temperature and humidity, and it may not be as efficient in humid environments.
- Water availability: Air cooled condensers do not require a water source, but water cooled condensers may be a more efficient choice in applications where a water source is readily available.
- Size and capacity: Air cooled condensers are generally limited in size and capacity compared to water cooled condensers, and may not be suitable for large or high heat load applications.

What are the limitations of using an air cooled condenser in a food and beverage processing facility?

There are a few limitations to using an air cooled condenser in a food and beverage processing facility:

- Operating temperature range: Air cooled condensers are generally limited to a certain operating temperature range, and may not be suitable for use in extremely cold or hot environments.
- Ambient temperature and humidity: The performance of an air cooled condenser can be affected by the ambient temperature and humidity, and it may not be as efficient in humid environments.
- Water availability: Air cooled condensers do not require a water source, but water cooled condensers may be a more efficient choice in applications where a water source is readily available.
- Size and capacity: Air cooled condensers are generally limited in size and capacity compared to water cooled condensers, and may not be suitable for large or high heat load applications.

What are the limitations of using an air cooled condenser in an HVAC system?

There are a few limitations to using an air cooled condenser in an HVAC system:

- Operating temperature range: Air cooled condensers are generally limited to a certain operating temperature range, and may not be suitable for use in extremely cold or hot environments.
- Ambient temperature and humidity: The performance of an air cooled condenser can be affected by the ambient temperature and humidity, and it may not be as efficient in humid environments.
- Water availability: Air cooled condensers do not require a water source, but water cooled condensers may be a more efficient choice in applications where a water source is readily available.
- Size and capacity: Air cooled condensers are generally limited in size and capacity compared to water cooled condensers, and may not be suitable for large or high heat load applications.

What are the main factors that affect the performance of an air cooled condenser?

There are several factors that can affect the performance of an air cooled condenser, including:

- Ambient temperature and humidity: The performance of an air cooled condenser can be affected by the ambient temperature and humidity. In general, the cooler and drier the air, the more efficient the condenser will be.
- Air flow: The performance of an air cooled condenser depends on the flow of air over the tubes or coils. If the air flow is restricted or insufficient, the condenser may not be able to dissipate heat effectively.
- Refrigerant charge: The performance of an air cooled condenser can be affected by the refrigerant charge in the system. If the charge is too low, the condenser may not be able to dissipate heat effectively.

- Tube or coil size: The size of the tubes or coils in the condenser can affect its performance. Larger tubes or coils may be more efficient, but may also be more expensive and take up more space.
- Fan and motor: The performance of the condenser's fan and motor can also affect the overall performance of the unit. If the fan or motor is not functioning properly, the condenser may not be able to dissipate heat effectively.

What are the main factors that affect the lifespan of an air cooled condenser?

There are several factors that can affect the lifespan of an air cooled condenser, including:

- Maintenance: Regular maintenance, including cleaning and checking and replacing components as needed, can extend the lifespan of an air cooled condenser.
- Operating conditions: The operating conditions, including the ambient temperature and humidity, can affect the lifespan of an air cooled condenser. In general, operating the condenser within its design parameters will help to extend its lifespan.
- Refrigerant type: The type of refrigerant used in the system can affect the lifespan of an air cooled condenser. Some refrigerants are more corrosive than others, and may shorten the lifespan of the condenser.
- Quality of components: The quality of the components used in the condenser, including the tubes or coils and the fan and motor, can also affect the lifespan of the unit.

How can the performance of an air cooled condenser be improved?

There are several ways to improve the performance of an air cooled condenser:

- Increase the surface area of the tubes or coils: Increasing the surface area of the tubes or coils in the condenser can improve its ability to dissipate heat.
- Use a high-efficiency fan: Using a high-efficiency fan in the condenser can help to improve its performance by increasing the flow of air over the tubes or coils.
- Use a microchannel condenser: A microchannel condenser uses small tubes or channels to transfer heat, which can be more efficient than traditional tubes or coils.
- Install an air-to-air heat exchanger: An air-to-air heat exchanger can be used to transfer heat from the condenser to the outside air, improving the overall efficiency of the system.
- Regular maintenance: Regular maintenance, including cleaning the condenser and checking and replacing components as needed, can also help to improve its performance.

How can the lifespan of an air cooled condenser be extended?

There are several ways to extend the lifespan of an air cooled condenser:

- Regular maintenance: Regular maintenance, including cleaning the condenser and checking and replacing components as needed, can help to extend its lifespan.
- Use high-quality components: Using high-quality components, such as tubes or coils and a fan and motor, can help to extend the lifespan of an air cooled condenser.
- Operate the condenser within its design parameters: Operating the condenser within its design parameters, including the ambient temperature and humidity range, can help to extend its lifespan.
- Use the correct refrigerant: Using the correct refrigerant for the system can help to extend the lifespan of the condenser, as some refrigerants are more corrosive than others.
- Avoid overloading the condenser: Overloading the condenser by using it beyond its capacity can shorten its lifespan. It is

important to properly size the condenser for the application to ensure that it is not overworked.

How often should an air cooled condenser be maintained?

The frequency of maintenance for an air cooled condenser will depend on the specific needs of the system and the operating conditions. In general, it is recommended to perform regular maintenance, including cleaning the condenser and checking and replacing components as needed, at least once a year. Some manufacturers may recommend more frequent maintenance for certain systems or operating conditions. It is important to follow the manufacturer's recommendations for maintenance to ensure the best performance and lifespan of the condenser.

What are the most common problems with air cooled condensers and how can they be addressed?

Some of the most common problems with air cooled condensers include:

- Reduced airflow: If the airflow over the tubes or coils is restricted or insufficient, the condenser may not be able to dissipate heat effectively. This can be caused by a dirty or damaged fan, or by debris blocking the air intake. Cleaning the condenser and the fan, and checking for and removing any debris, can help to improve airflow.
- Refrigerant charge issues: If the refrigerant charge in the system is too low, the condenser may not be able to dissipate heat effectively. This can be caused by a leak in the system, or by an incorrect charge. Checking for leaks and adding refrigerant as needed can help to address this issue.
- Failed components: Components such as the tubes or coils, the fan and motor, or the controls may fail over time. Replacing failed components can help to improve the performance of the condenser.

What safety precautions should be taken when working on an air cooled condenser?

There are several safety precautions that should be taken when working on an air cooled condenser:

- Wear personal protective equipment (PPE): It is important to wear appropriate PPE, such as gloves, safety glasses, and protective clothing, when working on an air cooled condenser.
- Follow proper handling procedures for refrigerants: Refrigerants can be hazardous if not handled properly. It is important to follow proper handling procedures, including wearing gloves and protective glasses, and using approved refrigerant recovery equipment.
- Disconnect power: Before working on the condenser, it is important to disconnect power to the unit to prevent electrical shock.
- Follow manufacturer's instructions: It is important to follow the manufacturer's instructions and warnings when working on the condenser to ensure safety.

How do I choose the right size air cooled condenser for my application?

There are several factors to consider when choosing the right size air cooled condenser for your application:

- Heat load: The heat load is the amount of heat that needs to be dissipated by the condenser. It is important to properly size the condenser for the heat load of the application to ensure that it can effectively dissipate heat.
- Operating conditions: The operating conditions, including the ambient temperature and humidity, should be considered when choosing the size of the condenser.
- Refrigerant type: The type of refrigerant used in the system can affect the size of the condenser needed.

- Space constraints: The available space for the condenser should also be considered when choosing the size.

It is generally recommended to consult with a qualified HVAC technician or engineer to determine the proper size air cooled condenser for your application.

How do I install an air cooled condenser?

Installing an air cooled condenser generally involves the following steps:

- Choose an appropriate location for the condenser: The condenser should be installed in a location with good airflow and easy access for maintenance.
- Prepare the mounting surface: The mounting surface should be level and able to support the weight of the condenser.
- Install the condenser: Follow the manufacturer's instructions for installing the condenser, including attaching any brackets or mounting hardware.
- Connect the electrical supply: Connect the electrical supply to the condenser according to the manufacturer's.

How do I maintain an air cooled condenser?

There are several steps that can be taken to maintain an air cooled condenser:

- Clean the condenser: The condenser should be cleaned regularly to remove any dirt or debris that may be blocking airflow. This can be done by using a hose to wash off the fins or by using a brush to gently clean the fins.
- Check and replace the fan and motor: The fan and motor should be checked regularly for wear or damage. If necessary, they should be replaced to ensure the condenser is functioning properly.
- Check and replace the tubes or coils: The tubes or coils should be checked regularly for leaks or damage. If necessary, they

should be replaced to ensure the condenser is functioning properly.

- Check and replace controls: The controls should be checked regularly to ensure they are functioning properly. If necessary, they should be replaced to ensure the condenser is functioning properly.
- Check and replace other components: Other components, such as the refrigerant charge and the wiring, should be checked regularly and replaced if necessary.

It is generally recommended to consult with a qualified HVAC technician or engineer for more specific maintenance recommendations for your air cooled condenser.

How do I troubleshoot an air cooled condenser?

There are several steps that can be taken to troubleshoot an air cooled condenser:

- Check the power supply: Make sure the condenser is receiving power and that all connections are secure.
- Check the refrigerant charge: Check the refrigerant charge in the system to ensure it is at the proper level.
- Check the fan and motor: Check the fan and motor to ensure they are functioning properly.
- Check the tubes or coils: Check the tubes or coils for leaks or damage.
- Check the controls: Check the controls to ensure they are functioning properly.
- Check for other issues: Other issues, such as a clogged air filter or a blocked air intake, should also be checked.

If the problem persists after these checks, it may be necessary to consult with a qualified HVAC technician or engineer for further troubleshooting and repair.

How do I dispose of an old or damaged air cooled condenser?

Disposing of an old or damaged air cooled condenser involves several steps:

- Remove refrigerant: The refrigerant in the condenser should be removed and properly disposed of according to local regulations and guidelines. This should be done by a qualified HVAC technician or engineer.
- Disconnect power: The power supply to the condenser should be disconnected before attempting to remove it.
- Remove the condenser: The condenser should be carefully removed from its location and properly packaged for disposal.
- Dispose of the condenser: The condenser should be disposed of according to local regulations and guidelines. This may include recycling or proper disposal at a designated facility.

It is important to follow proper procedures for disposing of an old or damaged air cooled condenser to ensure the safety of those handling the unit and to protect the environment.

What are the advantages of using an air cooled condenser over a water cooled condenser?

There are several advantages to using an air cooled condenser over a water cooled condenser:

- No water source required: Air cooled condensers do not require a water source, which can be an advantage in applications where water is not readily available or where the cost of water treatment is high.
- Lower initial cost: Air cooled condensers generally have a lower initial cost compared to water cooled condensers.
- Lower maintenance cost: Air cooled condensers may have a lower maintenance cost compared to water cooled condensers, as they do not require water treatment or cooling towers.
- Compact size: Air cooled condensers are generally smaller in size compared to water cooled condensers, which can be an advantage in applications with space constraints.

What are the disadvantages of using an air cooled condenser over a water cooled condenser?

There are several disadvantages to using an air cooled condenser over a water cooled condenser:

- Lower efficiency: Air cooled condensers may be less efficient than water cooled condensers, particularly in hot and humid conditions.
- Limited operating temperature range: Air cooled condensers are generally limited to a certain operating temperature range, and may not be suitable for use in extremely cold or hot environments.
- Limited capacity: Air cooled condensers are generally limited in size and capacity compared to water cooled condensers, and may not be suitable for large or high heat load applications.

What are the main applications of air cooled condensers?

Air cooled condensers are used in a variety of applications, including:

- HVAC systems: Air cooled condensers are commonly used in residential and commercial HVAC systems to dissipate heat from the refrigerant.
- Refrigeration systems: Air cooled condensers are also used in refrigeration systems, such as in supermarkets and restaurants, to dissipate heat from the refrigerant.
- Industrial process cooling: Air cooled condensers are used in industrial process cooling applications to dissipate heat from process fluids or gases.
- Food and beverage processing: Air cooled condensers are used in food and beverage processing facilities to dissipate heat from process fluids or gases.

What are the main types of air cooled condensers?

There are several types of air cooled condensers, including:

- Tube and fin condensers: Tube and fin condensers use tubes with fins attached to dissipate heat.
- Microchannel condensers: Microchannel condensers use small tubes or channels to transfer heat, which can be more efficient than traditional tubes or fins.
- Air-to-air heat exchangers: Air-to-air heat exchangers transfer heat from the condenser to the outside air, improving the overall efficiency of the system.
- Evaporative condensers: Evaporative condensers use a combination of air and water to dissipate heat, and are generally more efficient than air cooled condensers in hot and humid conditions.

How do I compare the efficiency of different air cooled condensers?

There are several factors to consider when comparing the efficiency of different air cooled condensers:

- Energy efficiency ratio (EER): The EER is a measure of the cooling efficiency of an air cooled condenser. A higher EER indicates a more efficient condenser.
- Seasonal energy efficiency ratio (SEER): The SEER is a measure of the cooling efficiency of an air cooled condenser over an entire cooling season. A higher SEER indicates a more efficient condenser.
- Coefficient of performance (COP): The COP is a measure of the efficiency of a refrigeration system, including the condenser. A higher COP indicates a more efficient system.
- Operating conditions: The efficiency of an air cooled condenser may vary depending on the operating conditions, such as the ambient temperature and humidity. It is important to consider these conditions when comparing the efficiency of different condensers.

Comparing the EER, SEER, and COP ratings of different air cooled condensers can provide a good indication of their relative efficiency. It is also important to consider the specific operating conditions and requirements of the application to determine the most efficient condenser for the needs of the system.

How do I choose the right type of air cooled condenser for my application?

There are several factors to consider when choosing the right type of air cooled condenser for your application:

- Heat load: The heat load is the amount of heat that needs to be dissipated by the condenser. It is important to choose a condenser that is capable of handling the heat load of the application.
- Operating conditions: The operating conditions, including the ambient temperature and humidity, should be considered when choosing the type of condenser. Some types of condensers, such as evaporative condensers, may be more efficient in hot and humid conditions.
- Refrigerant type: The type of refrigerant used in the system can affect the type of condenser needed.
- Space constraints: The available space for the condenser should also be considered when choosing the type.

It is generally recommended to consult with a qualified HVAC technician or engineer to determine the most appropriate type of air cooled condenser for your application.

How do I compare the cost of different air cooled condensers?

There are several factors to consider when comparing the cost of different air cooled condensers:

- Purchase price: The purchase price of the condenser is an important factor to consider when comparing the cost.
- Installation cost: The cost of installation should also be considered, as it can vary depending on the complexity of the

installation and the location of the condenser.

- Maintenance cost: The maintenance cost of the condenser should be considered, as it can vary depending on the type of condenser and the operating conditions.
- Energy cost: The energy cost of operating the condenser should also be considered, as more efficient condensers may have lower energy costs.

Comparing the total cost of ownership, including the purchase price, installation cost, maintenance cost, and energy cost, can provide a good indication of the overall cost of different air cooled condensers.

How do I compare the lifespan of different air cooled condensers?

The lifespan of an air cooled condenser can vary depending on a number of factors, including:

- Quality of the condenser: High quality condensers may have a longer lifespan than lower quality condensers.
- Operating conditions: The operating conditions, including the ambient temperature and humidity, can affect the lifespan of the condenser.
- Maintenance: Proper maintenance can extend the lifespan of an air cooled condenser.
- Type of condenser: Different types of condensers may have different lifespans. For example, evaporative condensers may have a shorter lifespan compared to other types of condensers due to the presence of water in the system.

It can be challenging to accurately compare the lifespan of different air cooled condensers, as there are many variables that can affect their lifespan. It is generally recommended to consult with a qualified HVAC technician or engineer for more specific lifespan estimates for different condensers.

How do I compare the noise level of different air cooled condensers?

The noise level of an air cooled condenser can be affected by several factors, including the type of condenser, the size of the condenser, and the operating conditions. It is generally recommended to consult with the manufacturer or a qualified HVAC technician or engineer to obtain noise level data for different condensers. The decibel (dB) level is commonly used to measure the noise level of air cooled condensers. A lower dB level indicates a quieter condenser.

How do I compare the durability of different air cooled condensers?

The durability of an air cooled condenser can vary depending on a number of factors, including:

- Quality of the condenser: High quality condensers may be more durable than lower quality condensers.
- Operating conditions: The operating conditions, including the ambient temperature and humidity, can affect the durability of the condenser.
- Maintenance: Proper maintenance can increase the durability of an air cooled condenser.
- Type of condenser: Different types of condensers may have different levels of durability. For example, evaporative condensers may be less durable compared to other types of condensers due to the presence of water in the system.

It can be challenging to accurately compare the durability of different air cooled condensers, as there are many variables that can affect their lifespan. It is generally recommended to consult with a qualified HVAC technician or engineer for more specific durability estimates for different condensers.

How do I compare the ease of maintenance of different air cooled condensers?

The ease of maintenance of an air cooled condenser can vary depending on a number of factors, including:

- Accessibility of components: Condensers with easily accessible components may be easier to maintain than those with components that are difficult to access.
- Number and complexity of components: Condensers with fewer and simpler components may be easier to maintain than those with more complex components.
- Type of condenser: Different types of condensers may have different levels of ease of maintenance. For example, evaporative condensers may be more difficult to maintain due to the presence of water in the system.
- Maintenance requirements: Some condensers may require more frequent maintenance than others.

It is generally recommended to consult with a qualified HVAC technician or engineer to obtain more specific information on the ease of maintenance of different air cooled condensers.

How do I compare the warranty of different air cooled condensers?

The warranty of an air cooled condenser can vary depending on the manufacturer and the specific condenser model. It is generally recommended to compare the terms of the warranty, including the length of coverage, the types of components covered, and any exclusions or limitations. It is also recommended to read the fine print of the warranty carefully to understand the manufacturer's obligations and the consumer's rights under the warranty.

How do I compare the brand reputation of different air cooled condensers?

There are several ways to compare the brand reputation of different air cooled condensers:

- Online reviews: Online reviews from customers and experts can provide valuable insights into the brand reputation of different condensers.

- Industry recognition: Brands that have received industry recognition, such as awards or certifications, may have a good reputation.
- Customer service: The quality of customer service provided by the brand can also be an indicator of their reputation.
- Product reliability: The reliability of the products offered by the brand can also be a factor in their reputation.

Comparing the brand reputation of different air cooled condensers can provide valuable insights into the quality and reliability of the products offered by each brand.

How do I compare the installation requirements of different air cooled condensers?

The installation requirements of different air cooled condensers can vary depending on the specific condenser model and the application. It is generally recommended to consult with the manufacturer or a qualified HVAC technician or engineer to obtain more specific information on the installation requirements of different condensers. Factors to consider when comparing the installation requirements may include the size and weight of the condenser, the required mounting surface, and the electrical and refrigerant connections needed.

How do I compare the serviceability of different air cooled condensers?

The serviceability of an air cooled condenser refers to the ease with which the condenser can be serviced or maintained. Factors to consider when comparing the serviceability of different air cooled condensers may include:

- Accessibility of components: Condensers with easily accessible components may be easier to service than those with components that are difficult to access.
- Service and maintenance instructions: Condensers with clear and comprehensive service and maintenance instructions may

be easier to service.

- Availability of parts: The availability of parts for the condenser can affect its serviceability. Condensers with readily available parts may be easier to service than those with hard-to-find parts.
- Service network: The presence of a large service network for the condenser can also affect its serviceability. Condensers with a wide network of service providers may be easier to service than those with a limited network.

It is generally recommended to consult with the manufacturer or a qualified HVAC technician or engineer to obtain more specific information on the serviceability of different air cooled condensers.

How do I compare the performance of different air cooled condensers?

There are several factors to consider when comparing the performance of different air cooled condensers:

- Energy efficiency: The energy efficiency of the condenser can be compared using metrics such as the energy efficiency ratio (EER) or the seasonal energy efficiency ratio (SEER).
- Cooling capacity: The cooling capacity of the condenser should be sufficient to meet the needs of the application.
- Operating conditions: The performance of the condenser may vary depending on the operating conditions, such as the ambient temperature and humidity.
- Refrigerant type: The type of refrigerant used in the system can affect the performance of the condenser.

Comparing the energy efficiency, cooling capacity, operating conditions, and refrigerant type of different air cooled condensers can provide a good indication of their relative performance. It is also important to consider the specific requirements and operating conditions of the application when evaluating the performance of different condensers.

How do I compare the flexibility of different air cooled condensers?

The flexibility of an air cooled condenser refers to its ability to adapt to changing conditions or requirements. Factors to consider when comparing the flexibility of different air cooled condensers may include:

- Modularity: Condensers that are modular in design may be more flexible as they can be easily expanded or modified to meet changing needs.
- Operating range: The operating range of the condenser, including the temperature and humidity range, can affect its flexibility.
- Refrigerant compatibility: The compatibility of the condenser with different types of refrigerants can affect its flexibility.
- Customization options: The availability of customization options, such as different sizes or configurations, can also affect the flexibility of the condenser.

It is generally recommended to consult with the manufacturer or a qualified HVAC technician or engineer to obtain more specific information on the flexibility of different air cooled condensers.

How do I compare the safety of different air cooled condensers?

The safety of an air cooled condenser can be affected by several factors, including:

- Safety features: Condensers with safety features, such as pressure relief valves or high temperature shut-off, may be safer than those without such features.
- Quality of construction: High quality construction can increase the safety of the condenser.
- Certification: Condensers that have been certified by a recognized organization, such as UL or CSA, may be safer than those that are not certified.

- Serviceability: Condensers that are easy to service may be safer as they can be more easily maintained and repaired.

It is generally recommended to consult with the manufacturer or a qualified HVAC technician or engineer to obtain more specific information on the safety of different air cooled condensers.

How do I compare the environmental impact of different air cooled condensers?

The environmental impact of an air cooled condenser can be affected by several factors, including:

- Energy efficiency: More efficient condensers may have a lower environmental impact due to their reduced energy consumption.
- Refrigerant type: The type of refrigerant used in the condenser can affect its environmental impact. Some refrigerants, such as HFCs, have a higher global warming potential compared to others.
- Recycling and disposal: The recycling and disposal of the condenser and its components can also affect its environmental impact.
- Manufacturing process: The manufacturing process of the condenser can also affect its environmental impact.

It is generally recommended to consult with the manufacturer or a qualified HVAC technician or engineer to obtain more specific information on the environmental impact of different air cooled condensers.

How do I compare the warranty of different air cooled condenser brands?

The warranty of an air cooled condenser can vary depending on the manufacturer and the specific condenser model. It is generally recommended to compare the terms of the warranty, including the length of coverage, the types of components covered, and any exclusions or limitations. It is also recommended to read the fine print of the warranty carefully to understand the manufacturer's obligations and the consumer's rights under the warranty.

How do I compare the reputation of different air cooled condenser brands?

There are several ways to compare the reputation of different air cooled condenser brands:

- Online reviews: Online reviews from customers and experts can provide valuable insights into the reputation of different brands.
- Industry recognition: Brands that have received industry recognition, such as awards or certifications, may have a good reputation.
- Customer service: The quality of customer service provided by the brand can also be an indicator of their reputation.
- Product reliability: The reliability of the products offered by the brand can also be a factor in their reputation.

Comparing the reputation of different air cooled condenser brands can provide valuable insights into the quality and reliability of the products offered by each brand.

How do I compare the installation requirements of different air cooled condenser brands?

The installation requirements of different air cooled condensers can vary depending on the specific condenser model and the application. It is generally recommended to consult with the manufacturer or a qualified HVAC technician or engineer to obtain more specific information on the installation requirements of different condensers. Factors to consider when comparing the installation requirements may include the size and weight of the condenser, the required mounting surface, and the electrical and refrigerant connections needed.

How do I compare the serviceability of different air cooled condenser brands?

The serviceability of an air cooled condenser refers to the ease with which the condenser can be serviced or maintained. Factors to consider when comparing the serviceability of different air cooled condenser brands may include:

- Accessibility of components: Condensers with easily accessible components may be easier to service than those with components that are difficult to access.
- Service and maintenance instructions: Condensers with clear and comprehensive service and maintenance instructions may be easier to service.
- Availability of parts: The availability of parts for the condenser can affect its serviceability. Condensers with readily available parts may be easier to service than those with hard-to-find parts.
- Service network: The presence of a large service network for the condenser can also affect its serviceability. Condensers with a wide network of service providers may be easier to service than those with a limited network.

It is generally recommended to consult with the manufacturer or a qualified HVAC technician or engineer to obtain more specific information on the serviceability of different air cooled condenser brands.

How do I compare the performance of different air cooled condenser brands?

There are several factors to consider when comparing the performance of different air cooled condenser brands:

- Energy efficiency: The energy efficiency of the condenser can be compared using metrics such as the energy efficiency ratio (EER) or the seasonal energy efficiency ratio (SEER).
- Cooling capacity: The cooling capacity of the condenser should be sufficient to meet the needs of the application.
- Operating conditions: The performance of the condenser may vary depending on the operating conditions, such as the ambient temperature and humidity.

- Refrigerant type: The type of refrigerant used in the system can affect the performance of the condenser.

Comparing the energy efficiency, cooling capacity, operating conditions, and refrigerant type of different air cooled condenser brands can provide a good indication of their relative performance. It is also important to consider the specific requirements and operating conditions of the application when evaluating the performance of different condensers.

How do I compare the flexibility of different air cooled condenser brands?

The flexibility of an air cooled condenser refers to its ability to adapt to changing conditions or requirements. Factors to consider when comparing the flexibility of different air cooled condenser brands may include:

- Modularity: Condensers that are modular in design may be more flexible as they can be easily expanded or modified to meet changing needs.
- Operating range: The operating range of the condenser, including the temperature and humidity range, can affect its flexibility.
- Refrigerant compatibility: The compatibility of the condenser with different types of refrigerants can affect its flexibility.
- Customization options: The availability of customization options, such as different sizes or configurations, can also affect the flexibility of the condenser.

It is generally recommended to consult with the manufacturer or a qualified HVAC technician or engineer to obtain more specific information on the flexibility of different air cooled condenser brands.

How do I compare the safety of different air cooled condenser brands?

The safety of an air cooled condenser can be affected by several factors, including:

- Safety features: Condensers with safety features, such as pressure relief valves or high temperature shut-off, may be safer than those without such features.
- Quality of construction: High quality construction can increase the safety of the condenser.
- Certification: Condensers that have been certified by a recognized organization, such as UL or CSA, may be safer than those that are not certified.
- Serviceability: Condensers that are easy to service may be safer as they can be more easily maintained and repaired.

It is generally recommended to consult with the manufacturer or a qualified HVAC technician or engineer to obtain more specific information on the safety of different air cooled condenser brands.

How do I compare the environmental impact of different air cooled condenser brands?

The environmental impact of an air cooled condenser can be affected by several factors, including:

- Energy efficiency: More efficient condensers may have a lower environmental impact due to their reduced energy consumption.
- Refrigerant type: The type of refrigerant used in the condenser can affect its environmental impact. Some refrigerants, such as HFCs, have a higher global warming potential compared to others.
- Recycling and disposal: The recycling and disposal of the condenser and its components can also affect its environmental impact.
- Manufacturing process: The manufacturing process of the condenser can also affect its environmental impact.

It is generally recommended to consult with the manufacturer or a qualified HVAC technician or engineer to obtain more specific information on the environmental impact of different air cooled condenser brands.

How do I compare the cost of different air cooled condensers?

The cost of an air cooled condenser can vary depending on a number of factors, including the size of the condenser, the efficiency rating, the type of refrigerant used, the location and installation requirements, and the brand and model. It is generally recommended to obtain quotes from multiple suppliers to compare the cost of different air cooled condensers. It is also important to consider the long-term operating costs, such as energy consumption and maintenance expenses, when comparing the cost of different condensers.

How do I compare the size of different air cooled condensers?

The size of an air cooled condenser is an important consideration as it needs to be suitable for the space available and the cooling requirements of the application. When comparing the size of different air cooled condensers, it is generally recommended to consider the following factors:

- Cooling capacity: The cooling capacity of the condenser should be sufficient to meet the needs of the application.
- Physical dimensions: The physical dimensions of the condenser, including the width, height, and depth, should be appropriate for the installation location.
- Weight: The weight of the condenser should be taken into account when determining the mounting surface and any necessary structural support.
- Refrigerant lines: The size of the refrigerant lines required for the condenser should also be considered when comparing the size of different models.

It is generally recommended to consult with the manufacturer or a qualified HVAC technician or engineer to obtain more specific information on the size of different air cooled condensers.

How do I compare the efficiency of different air cooled condensers?

The efficiency of an air cooled condenser can be compared using metrics such as the energy efficiency ratio (EER) or the seasonal energy efficiency ratio (SEER). These metrics are used to measure the efficiency of the condenser in converting electricity into cooling capacity. Higher values for EER or SEER generally indicate higher efficiency. It is also important to consider the operating conditions and the type of refrigerant used in the system when comparing the efficiency of different air cooled condensers.

How do I compare the cooling capacity of different air cooled condensers?

The cooling capacity of an air cooled condenser is an important factor to consider as it needs to be sufficient to meet the needs of the application. The cooling capacity of different air cooled condensers can be compared by considering the following factors:

- BTU rating: The BTU (British Thermal Unit) rating of the condenser indicates the amount of heat that the condenser can remove from the air. A higher BTU rating generally indicates a higher cooling capacity.
- Operating conditions: The cooling capacity of the condenser may vary depending on the operating conditions, such as the ambient temperature and humidity.
- Refrigerant type: The type of refrigerant used in the system can also affect the cooling capacity.

How do I compare the noise level of different air cooled condensers?

The noise level of an air cooled condenser can be an important consideration, especially if the condenser is located in a residential or commercial area where noise may be a concern. The noise level of different air cooled condensers can be compared by considering the following factors:

- Noise level rating: Some condensers may have a noise level rating indicated in decibels (dB). A lower dB rating generally indicates a lower noise level.
- Noise reduction features: Some condensers may have features designed to reduce noise, such as sound blankets or acoustic enclosures.
- Operating conditions: The noise level of the condenser may vary depending on the operating conditions, such as the speed of the fan and the temperature of the refrigerant.

It is generally recommended to consult with the manufacturer or a qualified HVAC technician or engineer to obtain more specific information on the noise level of different air cooled condensers.

How do I compare the durability of different air cooled condensers?

The durability of an air cooled condenser can be an important consideration, as a more durable condenser may have a longer lifespan and require fewer repairs. The durability of different air cooled condensers can be compared by considering the following factors:

- Quality of construction: The quality of the materials used in the construction of the condenser can affect its durability.
- Serviceability: Condensers that are easy to service may be more durable as they can be more easily maintained and repaired.
- Warranty: The length and terms of the warranty can also be an indicator of the durability of the condenser.

It is generally recommended to consult with the manufacturer or a qualified HVAC technician or engineer to obtain more specific information on the durability of different air cooled condensers.

How do I compare the maintenance requirements of different air cooled condensers?

The maintenance requirements of an air cooled condenser can vary depending on the specific condenser model and the operating conditions. It is generally recommended to consult with the manufacturer or a qualified HVAC technician or engineer to obtain more specific information on the maintenance requirements of different air cooled condensers. Factors to consider when comparing the maintenance requirements may include the frequency of maintenance tasks, such as cleaning and lubrication, and the availability of replacement parts.

How do I compare the installation costs of different air cooled condensers?

The installation costs of different air cooled condensers can vary depending on a number of factors, including the size and complexity of the installation, the location, and the availability of labor. It is generally recommended to obtain quotes from multiple suppliers or contractors to compare the installation costs of different air cooled condensers. It is also important to consider the long-term operating costs, such as energy consumption and maintenance expenses, when comparing the cost of different condensers.

How do I compare the operating costs of different air cooled condensers?

The operating costs of an air cooled condenser can include energy consumption, maintenance expenses, and repair costs. To compare the operating costs of different air cooled condensers, it is generally recommended to consider the following factors:

- Energy efficiency: More efficient condensers may have lower operating costs due to reduced energy consumption.
- Maintenance requirements: The frequency and cost of maintenance tasks can affect the operating costs of the condenser.
- Repair costs: The cost and frequency of repairs can also affect the operating costs of the condenser.
- Refrigerant type: The type of refrigerant used in the system can affect the operating costs of the condenser.

It is generally recommended to consult with the manufacturer or a qualified HVAC technician or engineer to obtain more specific information on the operating costs of different air cooled condensers.

How do I compare the reliability of different air cooled condensers?

The reliability of an air cooled condenser refers to its ability to operate consistently and effectively over time. To compare the reliability of different air cooled condensers, it is generally recommended to consider the following factors:

- Quality of construction: The quality of the materials used in the construction of the condenser can affect its reliability.
- Serviceability: Condensers that are easy to service may be more reliable as they can be more easily maintained and repaired.
- Warranty: The length and terms of the warranty can also be an indicator of the reliability of the condenser.
- Operating conditions: The reliability of the condenser may vary depending on the operating conditions, such as the ambient temperature and humidity.

It is generally recommended to consult with the manufacturer or a qualified HVAC technician or engineer to obtain more specific information on the reliability of different air cooled condensers.

How do I compare the versatility of different air cooled condensers?

The versatility of an air cooled condenser refers to its ability to be used in a variety of applications and environments. To compare the versatility of different air cooled condensers, it is generally recommended to consider the following factors:

- Operating range: The operating range of the condenser, including the temperature and humidity range, can affect its versatility.

- Refrigerant compatibility: The compatibility of the condenser with different types of refrigerants can affect its versatility.
- Customization options: The availability of customization options, such as different sizes or configurations, can also affect the versatility of the condenser.
- Installation requirements: The installation requirements of the condenser, such as the required mounting surface and electrical and refrigerant connections, can affect its versatility.

It is generally recommended to consult with the manufacturer or a qualified HVAC technician or engineer to obtain more specific information on the versatility of different air cooled condensers.

How do I compare the warranties of different air cooled condensers?

The warranty of an air cooled condenser can provide protection against defects in materials or workmanship and can be an important factor to consider when comparing different models. To compare the warranties of different air cooled condensers, it is generally recommended to consider the following factors:

- Length of warranty: The length of the warranty can vary depending on the manufacturer and model. A longer warranty may provide more protection against defects and failures.
- Coverage: The coverage of the warranty should be carefully reviewed to understand what is and is not covered. Some warranties may only cover certain components or defects, while others may provide more comprehensive coverage.
- Terms and conditions: The terms and conditions of the warranty, such as the requirement for regular maintenance or exclusions, should be carefully reviewed to understand the limitations and obligations of the warranty.
- Transferability: Some warranties may be transferable to a new owner if the condenser is sold or transferred, while others may not be.

It is generally recommended to consult with the manufacturer or a qualified HVAC technician or engineer to obtain more specific information on the warranties of different air cooled condensers.

How do I compare the customer support of different air cooled condenser brands?

The customer support of an air cooled condenser manufacturer can be an important factor to consider when comparing different brands. To compare the customer support of different air cooled condenser brands, it is generally recommended to consider the following factors:

- Availability: The availability of customer support, such as through phone or email, can affect the ability to obtain assistance or information.
- Response time: The response time of customer support may vary depending on the manufacturer and the availability of support staff.
- Knowledge and expertise: The knowledge and expertise of customer support staff can affect the quality of the assistance provided.
- Problem resolution: The ability of the customer support staff to resolve problems or issues may vary depending on the manufacturer.

It is generally recommended to consult with the manufacturer or a qualified HVAC technician or engineer to obtain more specific information on the customer support of different air cooled condenser brands.

How do I compare the reputation of different air cooled condenser brands?

The reputation of an air cooled condenser manufacturer can be an important factor to consider when comparing different brands. To compare the reputation of different air cooled condenser brands, it is generally recommended to consider the following factors:

- Customer reviews: Customer reviews and ratings can provide insight into the experiences of others with the brand and can help to identify common issues or strengths.
- Industry recognition: Industry recognition, such as awards or certifications, can be an indicator of the quality and reputation of the brand.
- Length of time in business: A brand that has been in business for a longer period of time may have a more established reputation.
- Market share: The market share of a brand can be an indicator of its popularity and reputation.

It is generally recommended to consult with the manufacturer or a qualified HVAC technician or engineer to obtain more specific information on the reputation of different air cooled condenser brands.

How do I compare the value of different air cooled condensers?

The value of an air cooled condenser can be a subjective determination and may depend on individual priorities and needs. To compare the value of different air cooled condensers, it is generally recommended to consider the following factors:

- Cost: The cost of the condenser should be considered in relation to the features, performance, and other factors offered.
- Energy efficiency: More efficient condensers may offer greater value due to reduced operating costs.
- Maintenance requirements: Condensers with lower maintenance requirements may offer greater value due to reduced maintenance costs.
- Warranty: The length and terms of the warranty can also be an indicator of the value of the condenser.
- Reputation: The reputation of the manufacturer and brand can also affect the perceived value of the condenser.

It is generally recommended to consult with the manufacturer or a qualified HVAC technician or engineer to obtain more specific information on the value of different air cooled condensers.

How do I compare the environmental impact of different air cooled condensers?

The environmental impact of an air cooled condenser can be an important consideration for some users. To compare the environmental impact of different air cooled condensers, it is generally recommended to consider the following factors:

- Energy efficiency: More efficient condensers may have a lower environmental impact due to reduced energy consumption.
- Refrigerant type: The type of refrigerant used in the system can affect the environmental impact of the condenser. Some refrigerants, such as chlorofluorocarbons (CFCs) and hydrofluorocarbons (HFCs), have been linked to ozone depletion and climate change.
- Recycling and disposal: The recycling and disposal of the condenser and its components can also affect the environmental impact.

It is generally recommended to consult with the manufacturer or a qualified HVAC technician or engineer to obtain more specific information on the environmental impact of different air cooled condensers.

How do I compare the aesthetics of different air cooled condensers?

The aesthetics of an air cooled condenser can be an important consideration, especially if the condenser is located in an area that is visible to the public. To compare the aesthetics of different air cooled condensers, it is generally recommended to consider the following factors:

- Size and shape: The size and shape of the condenser can affect its appearance.
- Color and finish: The color and finish of the condenser can also affect its appearance.

- Integration with surroundings: The ability of the condenser to blend in with or complement the surrounding environment may be an important consideration.

It is generally recommended to consult with the manufacturer or a qualified HVAC technician or engineer to obtain more specific information on the aesthetics of different air cooled condensers.

What is HVAC air-cooled condenser?

HVAC air-cooled condenser is a type of condenser used in heating, ventilation, and air conditioning (HVAC) systems. It is used to reject heat from the refrigerant, which is then expelled to the atmosphere. An air-cooled condenser consists of a series of finned tubes, through which the refrigerant flows, and fans that circulate air over the tubes to dissipate the heat. Unlike a water-cooled condenser, which uses a separate water source to transfer heat, an air-cooled condenser relies solely on air to transfer heat away from the refrigerant. This makes air-cooled condensers a popular choice in HVAC systems, especially in locations where water is scarce or where there are restrictions on the use of water for cooling.

CHAPTER TEN

X. Conclusion

In conclusion, air cooled condensers (ACCs) are a key technology in a variety of systems that rely on refrigeration or air conditioning, and they have the potential to play a significant role in the transition to a more sustainable and renewable energy system. There are a number of trends in ACC technology that are focused on improving energy efficiency, reducing greenhouse gas emissions, and increasing the use of low-GWP refrigerants. These trends include the use of advanced materials, improved fin design, variable speed drives, optimized air flow design, and the development of microchannel technology. By adopting these technologies, ACCs can help to reduce energy consumption, improve the overall efficiency of systems that use them, and contribute to the transition to a more sustainable energy system.

Summary of key points covered in the book

Reflection on the importance and versatility of ACCs in modern refrigeration and air conditioning systems.

However, air cooled condensers (ACCs) are an important and versatile technology in modern refrigeration and air conditioning systems. They are used to cool refrigerants or other fluids by dissipating heat to the air, and they are found in a wide range of applications, including refrigerated warehouses, supermarkets, and air conditioning systems in buildings.

ACCs can be designed to be more energy efficient, which can reduce the overall energy consumption of systems that use them. They can also be used in conjunction with other sustainable technologies, such as ground source heat pumps or solar thermal systems, to provide an efficient and environmentally friendly means of heating and cooling buildings. Additionally, the use of low-global warming potential (GWP) refrigerants in ACCs can contribute to the sustainability of energy systems by reducing the potential for these refrigerants to contribute to global warming when they are released into the atmosphere.

Disclaimer

Disclaimer for "Air-Cooled Condenser Fundamentals: Design, Operations, Troubleshooting, Maintenance, and Q&A":

This book is intended to provide general information about air-cooled condenser fundamentals, design, operations, troubleshooting, maintenance, and frequently asked questions (Q&A). The information in this book is provided with the understanding that the author and publisher are not engaged in rendering professional services. If professional advice or other expert assistance is required, the services of a competent professional person should be sought.

Although the author and publisher have taken reasonable care in preparing this book, they make no representation or warranties with respect to the accuracy or completeness of the contents of this book and specifically disclaim any implied warranties of merchantability or fitness for a particular purpose. No warranty may be created or extended by sales representatives or written sales materials. The advice and strategies contained herein may not be suitable for every situation.

The author and publisher shall not be liable for any loss of profit or any other commercial damages, including but not limited to special, incidental, consequential, or other damages. Readers should be aware that the information in this book may become outdated and that changes in the law, regulations, and technology can impact the accuracy and usefulness of the information in this book. Readers are encouraged to consult with appropriate professionals before taking any action based on the information in this book.

This book is not a substitute for professional training or certification in air-cooled condenser systems. The reader is solely responsible for any decisions or actions taken based on the information in this book.

This book is sold with the understanding that the publisher and author are not engaged in rendering legal, financial, or other professional advice. If legal, financial, or other expert assistance is required, the services of a competent professional person should be sought.

This disclaimer applies to the entire contents of this book.

Thank You

9 798889 597421

Printed by Libri Plureos GmbH in Hamburg,
Germany